A First Course in Electrical and Computer Engineering

with MATLAB™ Programs and Experiments

Louis L. Scharf
Richard T. Behrens

University of Colorado, Boulder

▲▼ **Addison-Wesley Publishing Company**

Reading, Massachusetts • Menlo Park, California • New York
Don Mills, Ontario • Wokingham, England • Amsterdam
Bonn • Sydney • Singapore • Tokyo • Madrid • San Juan

This book was produced by the Addison-Wesley Electronic Production Department on an Apple Macintosh IIcx from TEXtures files supplied by the authors. The output was generated by the American Mathematical Society.

Library of Congress Cataloging-in-Publication Data

Scharf, Louis L.
    A first course in electrical and computer engineering: with MATLAB programs and experiments / Louis L. Scharf, Richard T. Behrens.
        p.   cm.
        Includes index.
        ISBN 0-201-53472-X
        1. Electrical engineering—Problems, exercises, etc.   2. Computer engineering—Problems, exercises, etc.   3.   MATLAB (Computer program)
I. Behrens, Richard T.   II. Title.
    TK168.S34 1990
    621.3'076–dc20   90-888
                                                                                CIP

2 3 4 5 6 7 8 9–MA– 95 94 93 92 91

Louis Scharf dedicates this book to
his wife Carol, son Greg, and daughter Heidi,
for love and inspiration;
his parents Louis Sr. and Ann,
in celebration of their 50th wedding anniversary.

Richard Behrens dedicates this book to
his wife Debbie, and child as yet unborn,
for love and encouragement;
his parents Richard and Elsie,
in gratitude for a good education.

| | | | | | | | **Preface** |

This book was written for an experimental freshman course at the University of Colorado. The course is now an elective that the majority of our electrical and computer engineering students take in the second semester of their freshman year, just before their first circuits course. Our department decided to offer this course for several reasons:

(1) we wanted to pique students' interest in engineering by acquainting them with engineering teachers early in their university careers and by providing them with exposure to the types of problems that electrical and computer engineers are asked to solve;

(2) we wanted students entering the electrical and computer engineering programs to be prepared in complex analysis, phasors, and linear algebra, topics that are of fundamental importance in our discipline;

(3) we wanted students to have an introduction to a software application tool, such as MATLAB, to complete their preparation for practical and efficient computing in their subsequent courses and in their professional careers;

(4) we wanted students to make early contact with advanced topics like vector graphics, filtering, and binary coding so that they would gain a more rounded picture of modern electrical and computer engineering.

In order to introduce this course, we had to sacrifice a second semester of Pascal programming. We concluded that the sacrifice was worth making because we had found that most of our students were prepared for high-level language computing after just one semester of programming.

We believe engineering educators elsewhere are reaching similar conclusions about their own students and curriculums. We hope this book helps create a much needed dialogue about curriculum revision and that it leads to the development of similar introductory courses that encourage students to enter and practice our craft.

Students electing to take this course have completed one semester of calculus, computer programming, chemistry, and humanities. Concurrently with this course, students take physics and a second semester of calculus, as well as a second semester in the humanities. By omitting the advanced topics marked by asterisks, we are able to cover Chapters 1 through 4, plus two of the three remaining chapters. The book is organized so that the instructor can select any two of the three. If every chapter of this book is covered, including the advanced topics, then enough material exists for a two-semester course.

The first three chapters of this book provide a fairly complete coverage of complex numbers, the functions e^x and $e^{j\theta}$, and phasors. Our department philosophy is that these topics must be understood if a student is to succeed in electrical and computer engineering. These three chapters may also be used as a supplement to a circuits course. A measured pace of presentation, taking between sixteen and eighteen lectures, is sufficient to cover all but the advanced sections in Chapters 1 through 3.

Chapter 4 on linear algebra is prerequisite for all subsequent chapters. We use eight to ten lectures to cover it. We devote twelve to sixteen lectures to cover topics from Chapters 5 through 7. (We assume a semester consisting of 42 lectures and three exams.) Chapter 5 applies the linear algebra learned in the previous chapter to the problem of translating, scaling, and rotating images. Chapter 6 introduces the student to basic ideas in averaging and filtering. Chapter 7 covers the rudiments of binary coding, including Huffman codes and Hamming codes.

If the users of this book find Chapters 5 through 7 too confining, we encourage them to supplement the essential material in Chapters 1 through 4 with their own course notes on additional topics. Within electrical and computer engineering there are endless possibilities. Practically any set of topics that can be taught with conviction and enthusiasm will whet the student's appetite. We encourage you to write to us or to our editor, Tom Robbins, about your ideas for additional topics. We would like to think that our book and its subsequent editions will have an open architecture that enables us to accommodate a wide range of student and faculty interests.

Throughout this book we have used MATLAB programs to illustrate key ideas. MATLAB is an interactive, matrix-oriented language that is ideally suited to circuit analysis, linear systems, control theory, communications, linear algebra, and numerical analysis. MATLAB is rapidly becoming a standard software tool in universities and engineering companies. (For more information about MATLAB, return the attached card in the back of this book to The MathWorks, Inc.) MATLAB programs are designed to develop the student's ability to solve meaningful problems, compute, and plot in a high-level applications language. Our students get started in MATLAB by working through Appendix 1, "An Introduction to MATLAB," while seated at an IBM PC (or look-alike) or an Apple Macintosh. We also have them run through the demonstration programs in Chapter 1. Each week we give three classroom lectures and conduct a one-hour computer lab session. Students use this lab session to hone MATLAB skills, to write programs, or to conduct the numerical experiments that are given at the end of each chapter. We require that these experiments be carried out and then reported in a short lab report that contains (i) introduction, (ii) analytical computations, (iii) computer code, (iv) experimental results, and (v) conclusions. The quality of the numerical results and the computer graphics astonishes students. Solutions to the chapter problems are available from the publisher for instructors who adopt this text for classroom use.

We wish to acknowledge our late colleague Richard Roberts, who encouraged us to publish this book, and Michael Lightner and Ruth Ravenel, who taught Chapters 4 and 5 and offered helpful suggestions on the manuscript. We thank C. T. Mullis for allowing us to use his notes on binary codes to guide our writing of Chapter 7. We thank Cédric Demeure and Peter Massey for their contributions to the writing of Appendix 1, "An Introduction to MATLAB," and Appendix 2, "The Edix Editor." We thank Tom Robbins, our editor at Addison-Wesley, for his encouragement, patience, and many suggestions. We are especially grateful to Julie Fredlund, who composed this text through many drafts and improved it in many ways. We thank her for preparing an excellent manuscript for production.

L. L. Scharf

R. T. Behrens

Boulder, Colorado

To the Teacher:

An incomplete understanding of complex numbers and phasors handicaps students in circuits and electronics courses, and even more so in advanced courses such as electromagnetics, optics, linear systems, control, and communication systems. Our faculty has decided to address this problem as early as possible in the curriculum by designing a course that drills complex numbers and phasors into the minds of beginning engineering students. We have used power signals, musical tones, Lissajous figures, light scattering, and RLC circuits to illustrate the usefulness of phasor calculus. Chapters 4 through 7 introduce students to a handful of modern ideas in electrical and computer engineering. The motivation is to whet students' appetites for more advanced problems. The topics we have chosen—linear algebra, vector graphics, filtering, and binary codes—are only representative.

At first glance, many of the equations in this book look intimidating to beginning students. For this reason, we proceed at a very measured pace. In our lectures, we write out in agonizing detail every equation that involves a sequence or series. For example, the sum $\sum_{n=0}^{N-1} z^n$ is written out as

$$1 + z + z^2 + \cdots + z^{N-1},$$

and then it is evaluated for some specific value of z before we derive the analytical result $\frac{1-z^N}{1-z}$. Similarly, an infinite sequence like $\lim_{n \to \infty} \left(1 + \frac{x}{n}\right)^n$ is written out as

$$(1+x), \left(1 + \frac{x}{2}\right)^2, \left(1 + \frac{x}{3}\right)^3, \ldots, \left(1 + \frac{x}{100}\right)^{100}, \ldots,$$

and then it is evaluated for some specific x and for several values of n before the limit is derived. We try to preserve this practice of pedantic excess until it is clear that every student is comfortable with an idea and the notation for coding the idea.

To the Student:

These are exciting times for electrical and computer engineering. To celebrate its silver anniversary, the National Academy of Engineering announced in February of 1990 the top ten engineering feats of the previous twenty-five years. The Apollo moon landing, a truly Olympian and protean achievement, ranked number one. However, a number of other achievements in the top ten were also readily identifiable as the products of electrical and computer engineers:

(1) communication and remote sensing satellites,

(2) the microprocessor,

(3) computer-aided design and manufacturing (CADCAM),

(4) computerized axial tomography (CAT scan),

(5) lasers, and

(6) fiber optic communication.

As engineering students, you recognize these achievements to be important milestones for humanity; you take pride in the role that engineers have played in the technological revolution of the twentieth century.

So how do we harness your enthusiasm for the grand enterprise of engineering? Historically, we have enrolled you in a freshman curriculum of mathematics, science, and humanities. If you succeeded, we enrolled you in an engineering curriculum. We then taught you the details of your profession and encouraged your faith that what you were studying is what you must study to be creative and productive engineers. The longer your faith held, the more likely you were to complete your studies. This seems like an imperious approach to engineering education, even though mathematics, physics, and the humanities are the foundation of engineering, and details are what form the structure of engineering. It seems to us that a better way to stimulate your enthusiasm and encourage your faith is to introduce you early in your studies to engineering teachers who will share their insights about some of the fascinating advanced topics in engineering, while teaching you the mathematical and physical principles of engineering. But you must match the teacher's commitment with your own commitment to study. This means that you must attend lectures, read texts, and work problems. You must be inquisitive and skeptical. Ask yourself how an idea is limited in scope and how it might be extended to apply to a wider range of problems. For, after all, one of the great themes of engineering is that a few fundamental ideas from mathematics and science, coupled with a few principles of design, may be applied to a wide range of engineering problems. Good luck with your studies.

| | | | | | | **Contents** |

1

Complex Numbers

Notes to Teachers and Students:

When we teach complex numbers to beginning engineering students, we encourage a *geometrical picture* supported by an *algebraic structure*. Every algebraic manipulation carried out in a lecture is accompanied by a carefully drawn picture in order to fix the idea that geometry and algebra go hand-in-glove to complete our understanding of complex numbers. We assign essentially every problem for homework.

We use the MATLAB programs in this chapter to illustrate the theory of complex numbers and to develop skill with the MATLAB language. The numerical experiment in Section 1.7 introduces students to *the* basic quadratic equation of electrical and computer engineering and shows how the roots of this quadratic equation depend on the coefficients of the equation.

Section 1.5, "Representing Complex Numbers in a Vector Space," is a little demanding for freshmen but easily accessible to sophomores. It may be covered for additional insight, skipped without consequence, or covered after Chapter 4. Section 1.6, "An Electric Field Computation," is well beyond most freshmen, and it is demanding for sophomores. Nonetheless, an expert in electromagnetics might want to cover Section 1.6 for the insight it brings to the use of complex numbers for representing two-dimensional real quantities.

1.1 Introduction

It is hard to overestimate the value of complex numbers. They first arose in the study of roots for quadratic equations. But, as with so many other great discoveries, complex numbers have found widespread application well outside their original domain of discovery. They are now used throughout mathematics, applied science, and engineering to represent the harmonic nature of vibrating systems and oscillating fields. For example, complex numbers may be used to study

(i) traveling waves on a sea surface;

(ii) standing waves on a violin string;

(iii) the pure tone of a Kurzweil piano;

(iv) the acoustic field in a concert hall;

(v) the light of a He-Ne laser;

(vi) the electromagnetic field in a light show;

(vii) the vibrations in a robot arm;

(viii) the oscillations of a suspension system;

(ix) the carrier signal used to transmit AM or FM radio;

(x) the carrier signal used to transmit digital data over telephone lines; and

(xi) the 60 Hz signal used to deliver power to a home.

In this chapter we develop the algebra and geometry of complex numbers. In Chapter 3 we will show how complex numbers are used to build phasor representations of power and communication signals.

1.2 Geometry of Complex Numbers

The most fundamental new idea in the study of complex numbers is the "imaginary number" j. This imaginary number is defined to be the square root of -1:

$$j = \sqrt{-1} \qquad\qquad 1.1$$

$$j^2 = -1.$$

The imaginary number j is used to build complex numbers z from real numbers x and y in the following way:

$$z = x + jy. \qquad\qquad 1.2$$

We say that the complex number z has "real part" x and "imaginary part" y:

$$z = \text{Re}[z] + j\,\text{Im}[z] \qquad\qquad 1.3$$

$$\text{Re}[z] = x; \qquad \text{Im}[z] = y.$$

In MATLAB, the variable x is denoted by `real(z)`, and the variable y is denoted by `imag(z)`. In communication theory, x is called the "in-phase" component of z, and y is called the "quadrature" component. We call $z = x + jy$ the *Cartesian* representation of z, with real component x and imaginary component y. We say that the Cartesian pair (x, y) *codes* the complex number z.

We may plot the complex number z on the plane as in Figure 1.1. We call the horizontal axis the "real axis" and the vertical axis the "imaginary axis." The plane is called the "complex plane." The radius and angle of the line to the point $z = x + jy$ are

$$r = \sqrt{x^2 + y^2} \qquad 1.4$$

$$\theta = \tan^{-1}\left(\frac{y}{x}\right).$$

See Figure 1.1. In MATLAB, r is denoted by `abs(z)`, and θ is denoted by `angle(z)`.

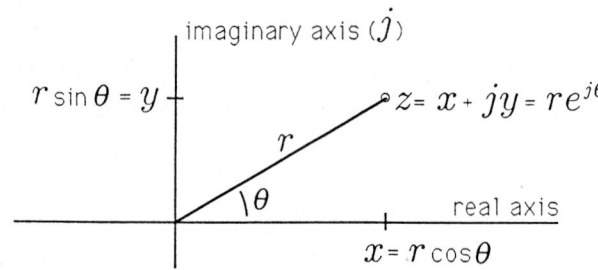

Figure 1.1: Cartesian and Polar Representations of the Complex Number z

The original Cartesian representation is obtained from the radius r and angle θ as follows:

$$x = r \cos \theta \qquad 1.5$$

$$y = r \sin \theta.$$

The complex number z may therefore be written as

$$z = x + jy$$
$$= r \cos \theta + jr \sin \theta \qquad 1.6$$
$$= r(\cos \theta + j \sin \theta).$$

The complex number $\cos\theta + j\sin\theta$ is, itself, a number that may be represented on the complex plane and coded with the Cartesian pair $(\cos\theta, \sin\theta)$. This is illustrated in Figure 1.2. The radius and angle to the point $z = \cos\theta + j\sin\theta$ are 1 and θ. Can you see why?

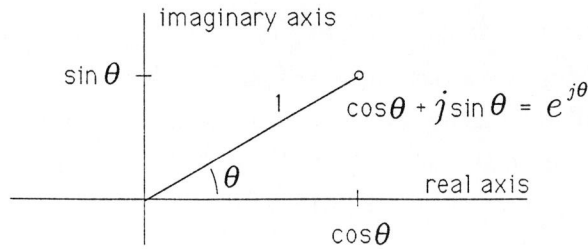

Figure 1.2: The Complex Number $\cos\theta + j\sin\theta$

The complex number $\cos\theta + j\sin\theta$ is of such fundamental importance to our study of complex numbers that we give it the special symbol $e^{j\theta}$:

$$e^{j\theta} = \cos\theta + j\sin\theta. \qquad 1.7$$

As illustrated in Figure 1.2, the complex number $e^{j\theta}$ has radius 1 and angle θ. With the symbol $e^{j\theta}$, we may write the complex number z as

$$z = re^{j\theta}. \qquad 1.8$$

We call $z = re^{j\theta}$ a *polar* representation for the complex number z. We say that the polar pair $r \angle\theta$ *codes* the complex number z. In this polar representation, we define $|z| = r$ to be the *magnitude* of z and $\arg(z) = \theta$ to be the angle, or *phase*, of z:

$$|z| = r \qquad 1.9$$

$$\arg(z) = \theta.$$

With these definitions of magnitude and phase, we can write the complex number z as

$$z = |z|e^{j\,\arg(z)}. \qquad 1.10$$

Let's summarize our ways of writing the complex number z and record the corresponding geometric codes:

$$z = x + jy = re^{j\theta} = |z|e^{j \arg(z)}.$$

$\downarrow$ $\downarrow$

(x, y) $r \angle \theta$

1.11

In Section 1.4 we show that the definition $e^{j\theta} = \cos\theta + j\sin\theta$ is more than symbolic. We show, in fact, that $e^{j\theta}$ is just the familiar function e^x evaluated at the imaginary argument $x = j\theta$. We call $e^{j\theta}$ a "complex exponential," meaning that it is an exponential with an imaginary argument.

Problem 1.1 Prove $(j)^{2n} = (-1)^n$ and $(j)^{2n+1} = (-1)^n j$. Evaluate j^3, j^4, j^5. ∎

Problem 1.2 Prove $e^{j[(\pi/2)+m2\pi]} = j$, $e^{j[(3\pi/2)+m2\pi]} = -j$, $e^{j(0+m2\pi)} = 1$, and $e^{j(\pi+m2\pi)} = -1$. Plot these identities on the complex plane. (Assume m is an integer.) ∎

Problem 1.3 Find the polar representation $z = re^{j\theta}$ for each of the following complex numbers:
  (a) $z = 1 + j0$;
  (b) $z = 0 + j1$;
  (c) $z = 1 + j1$;
  (d) $z = -1 - j1$.
Plot the points on the complex plane. ∎

Problem 1.4 Find the Cartesian representation $z = x + jy$ for each of the following complex numbers:
  (a) $z = \sqrt{2}\, e^{j\pi/2}$;
  (b) $z = \sqrt{2}\, e^{j\pi/4}$;
  (c) $z = e^{j3\pi/4}$;
  (d) $z = \sqrt{2}\, e^{j3\pi/2}$.
Plot the points on the complex plane. ∎

Problem 1.5 The following geometric codes represent complex numbers. Decode each by writing down the corresponding complex number z:

  (a) $(0.7, -0.1)$ $z = ?$

  (b) $(-1.0, 0.5)$ $z = ?$

  (c) $1.6 \angle \pi/8$ $z = ?$

  (d) $0.4 \angle 7\pi/8$ $z = ?$ ■

Problem 1.6 Show that $\mathrm{Im}[jz] = \mathrm{Re}[z]$ and $\mathrm{Re}[-jz] = \mathrm{Im}[z]$. ■

Demo 1.1 (MATLAB). Run the following MATLAB program in order to compute and plot the complex number $e^{j\theta}$ for $\theta = i2\pi/360$, $i = 1, 2, \ldots, 360$:

```
j=sqrt(-1)
n=360
for i=1:n,circle(i)=exp(j*2*pi*i/n);end;
axis('square')
plot(circle)
```

Replace the explicit **for** loop of line 3 by the implicit loop

```
circle=exp(j*2*pi*[1:n]/n);
```

to speed up the calculation. You can see from Figure 1.3 that the complex number $e^{j\theta}$, evaluated at angles $\theta = 2\pi/360, 2(2\pi/360), \ldots$, turns out complex numbers that lie at angle θ and radius 1. We say that $e^{j\theta}$ is a complex number that lies on the unit circle. We will have much more to say about the unit circle in Chapter 2. □

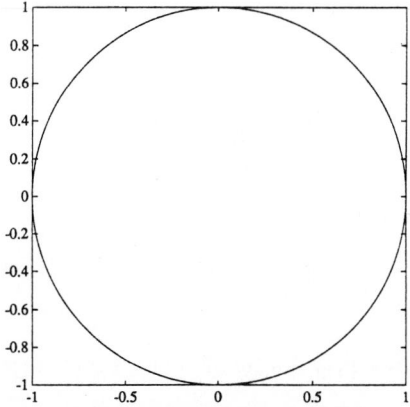

Figure 1.3: The Complex Numbers $e^{j\theta}$ for $0 \le \theta \le 2\pi$ (**Demo 1.1**)

1.3 Algebra of Complex Numbers

The complex numbers form a mathematical "field" on which the usual operations of addition and multiplication are defined. Each of these operations has a simple geometric interpretation.

Addition and Multiplication. The complex numbers z_1 and z_2 are added according to the rule

$$z_1 + z_2 = (x_1 + jy_1) + (x_2 + jy_2)$$
$$= (x_1 + x_2) + j(y_1 + y_2).$$

1.12

We say that the real parts add and the imaginary parts add. As illustrated in Figure 1.4, the complex number $z_1 + z_2$ is computed from a "parallelogram rule," wherein $z_1 + z_2$ lies on the node of a parallelogram formed from z_1 and z_2.

Problem 1.7 Let $z_1 = r_1 e^{j\theta_1}$ and $z_2 = r_2 e^{j\theta_2}$. Find a polar formula $z_3 = r_3 e^{j\theta_3}$ for $z_3 = z_1 + z_2$ that involves *only* the variables r_1, r_2, θ_1, and θ_2. The formula for r_3 is the "law of cosines." ∎

The product of z_1 and z_2 is

$$z_1 z_2 = (x_1 + jy_1)(x_2 + jy_2)$$
$$= (x_1 x_2 - y_1 y_2) + j(y_1 x_2 + x_1 y_2).$$

1.13

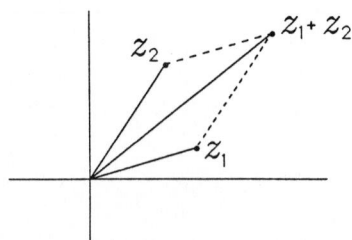

Figure 1.4: Adding Complex Numbers

If the polar representations for z_1 and z_2 are used, then the product may be written as[1]

$$
\begin{aligned}
z_1 z_2 &= r_1 e^{j\theta_1} r_2 e^{j\theta_2} \\
&= (r_1 \cos\theta_1 + jr_1 \sin\theta_1)(r_2 \cos\theta_2 + jr_2 \sin\theta_2) \\
&= (r_1 \cos\theta_1 r_2 \cos\theta_2 - r_1 \sin\theta_1 r_2 \sin\theta_2) \\
&\quad + j(r_1 \sin\theta_1 r_2 \cos\theta_2 + r_1 \cos\theta_1 r_2 \sin\theta_2) \\
&= r_1 r_2 \cos(\theta_1 + \theta_2) + jr_1 r_2 \sin(\theta_1 + \theta_2) \\
&= r_1 r_2 e^{j(\theta_1 + \theta_2)}.
\end{aligned}
\qquad 1.14
$$

We say that the magnitudes multiply and the angles add. As illustrated in Figure 1.5, the product $z_1 z_2$ lies at the angle $(\theta_1 + \theta_2)$.

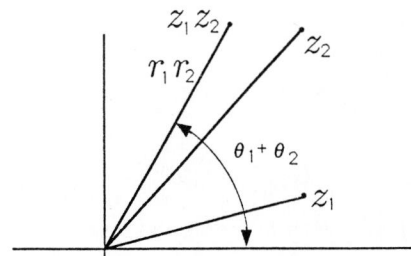

Figure 1.5: Multiplying Complex Numbers

Rotation. There is a special case of complex multiplication that will become very important in our study of phasors in Chapter 3. When z_1 is the complex number $z_1 = r_1 e^{j\theta_1}$ and z_2 is the complex number $z_2 = e^{j\theta_2}$, then the product of z_1 and z_2 is

$$
z_1 z_2 = z_1 e^{j\theta_2} = r_1 e^{j(\theta_1 + \theta_2)}.
\qquad 1.15
$$

As illustrated in Figure 1.6, $z_1 z_2$ is just a rotation of z_1 through the angle θ_2.

---

[1] We have used the trigonometric identities $\cos(\theta_1 + \theta_2) = \cos\theta_1 \cos\theta_2 - \sin\theta_1 \sin\theta_2$ and $\sin(\theta_1 + \theta_2) = \sin\theta_1 \cos\theta_2 + \cos\theta_1 \sin\theta_2$ to derive this result.

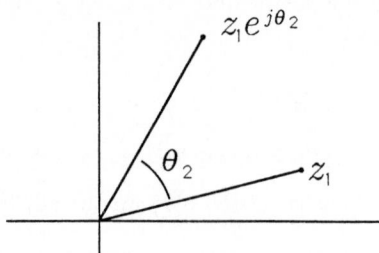

Figure 1.6: Rotation of Complex Numbers

Problem 1.8   Begin with the complex number $z_1 = x + jy = re^{j\theta}$. Compute the complex number $z_2 = jz_1$ in its Cartesian and polar forms. The complex number z_2 is sometimes called perp(z_1). Explain why by writing perp(z_1) as $z_1 e^{j\theta_2}$. What is θ_2? Repeat this problem for $z_3 = -jz_1$. ∎

Powers. If the complex number z_1 multiplies itself N times, then the result is

$$(z_1)^N = r_1^N e^{jN\theta_1}. \tag{1.16}$$

This result may be proved with a simple induction argument. Assume $z_1^k = r_1^k e^{jk\theta_1}$. (The assumption is true for $k = 1$.) Then use the recursion $z_1^{k+1} = z_1^k z_1 = r_1^{k+1} e^{j(k+1)\theta_1}$. Iterate this recursion (or induction) until $k + 1 = N$. Can you see that, as n ranges from $n = 1, \ldots, N$, the angle of z_1^n ranges from θ_1 to $2\theta_1, \ldots,$ to $N\theta_1$ and the radius ranges from r_1 to $r_1^2, \ldots,$ to r_1^N? This result is explored more fully in Problem 1.19.

Complex Conjugate. Corresponding to every complex number $z = x + jy = re^{j\theta}$ is the complex conjugate

$$z^* = x - jy = re^{-j\theta}. \tag{1.17}$$

The complex number z and its complex conjugate are illustrated in Figure 1.7. The recipe for finding complex conjugates is to "change j to $-j$." This changes the sign of the imaginary part of the complex number.

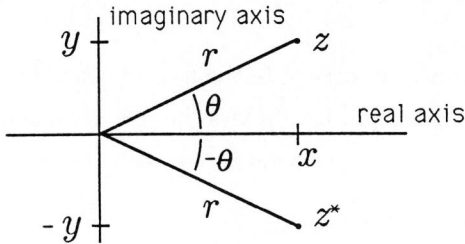

Figure 1.7: A Complex Variable and Its Complex Conjugate

Magnitude Squared. The product of z and its complex conjugate is called the *magnitude squared* of z and is denoted by $|z|^2$:

$$|z|^2 = z^*z = (x - jy)(x + jy) = x^2 + y^2 = re^{-j\theta}re^{j\theta} = r^2. \qquad 1.18$$

Note that $|z| = r$ is the radius, or magnitude, that we defined in Section 1.2.

Problem 1.9 Write z^* as $z^* = zw$. Find w in its Cartesian and polar forms. ∎

Problem 1.10 Prove that angle $(z_2 z_1^*) = \theta_2 - \theta_1$. ∎

Problem 1.11 Show that the real and imaginary parts of $z = x + jy$ may be written as

$$\text{Re}[z] = \frac{1}{2}(z + z^*)$$

$$\text{Im}[z] = \frac{1}{2j}(z - z^*). \quad \blacksquare$$

Commutativity, Associativity, and Distributivity. The complex numbers commute, associate, and distribute under addition and multiplication as follows:

$$z_1 + z_2 = z_2 + z_1 \qquad 1.19$$

$$z_1 z_2 = z_2 z_1$$

$$(z_1 + z_2) + z_3 = z_1 + (z_2 + z_3)$$

$$z_1(z_2 z_3) = (z_1 z_2)z_3$$

$$z_1(z_2 + z_3) = z_1 z_2 + z_1 z_3.$$

Identities and Inverses. In the field of complex numbers, the complex number $0 + j0$ (denoted by 0) plays the role of an additive identity, and the complex number $1 + j0$ (denoted by 1) plays the role of a multiplicative identity:

$$z + 0 = z = 0 + z \qquad\qquad 1.20$$

$$z1 = z = 1z.$$

In this field, the complex number $-z = -x + j(-y)$ is the additive inverse of z, and the complex number $z^{-1} = \frac{x}{x^2+y^2} + j\left(\frac{-y}{x^2+y^2}\right)$ is the multiplicative inverse:

$$z + (-z) = 0 \qquad\qquad 1.21$$

$$zz^{-1} = 1.$$

Problem 1.12 Show that the additive inverse of $z = re^{j\theta}$ may be written as $re^{j(\theta+\pi)}$. ∎

Problem 1.13 Show that the multiplicative inverse of z may be written as

$$z^{-1} = \frac{1}{z^* z} z^* = \frac{1}{x^2 + y^2}(x - jy).$$

Show that $z^* z$ is real. Show that z^{-1} may also be written as

$$z^{-1} = r^{-1}e^{-j\theta}.$$

Plot z and z^{-1} for a representative z. ∎

Problem 1.14 Prove $(j)^{-1} = -j$. ∎

Problem 1.15 Find z^{-1} when $z = 1 + j1$. ∎

Problem 1.16 Prove $(z^{-1})^* = (z^*)^{-1} = r^{-1}e^{j\theta} = \frac{1}{z^*z}\,z$. Plot z and $(z^{-1})^*$ for a representative z. ∎

Problem 1.17 Find all of the complex numbers z with the property that $jz = -z^*$. Illustrate these complex numbers on the complex plane. ∎

Demo 1.2 (MATLAB). Create and run the following script file (name it `Complex Numbers`)[2]

```
clear, clg
j=sqrt(-1)
z1=1+j*.5,z2=2+j*1.5
z3=z1+z2,z4=z1*z2
z5=conj(z1),z6=j*z2
axis([-4 4 -4 4]),axis('square'),plot(z1,'o')
hold on
plot(z2,'0'),plot(z3,'+'),plot(z4,'*'),
plot(z5,'x'),plot(z6,'x')
```

With the help of Appendix 1, you should be able to annotate each line of this program. View your graphics display to verify the rules for add, multiply, conjugate, and perp. See Figure 1.8. □

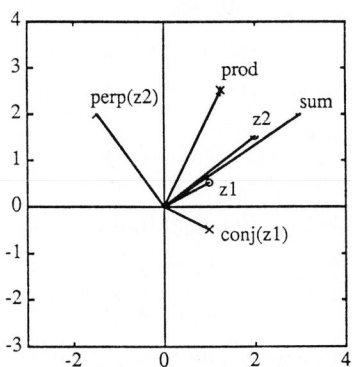

Figure 1.8: Complex Numbers **(Demo 1.2)**

---

[2] If you are using PC-MATLAB, you will need to name your file `cmplxnos.m`.

Problem 1.18   Prove that $z^0 = 1$. ∎

Problem 1.19   (MATLAB) Choose $z_1 = 1.05e^{j2\pi/16}$ and $z_2 = 0.95e^{j2\pi/16}$. Write a MATLAB program to compute and plot z_1^n and z_2^n for $n = 1, 2, \ldots, 32$. You should observe a figure like Figure 1.9. ∎

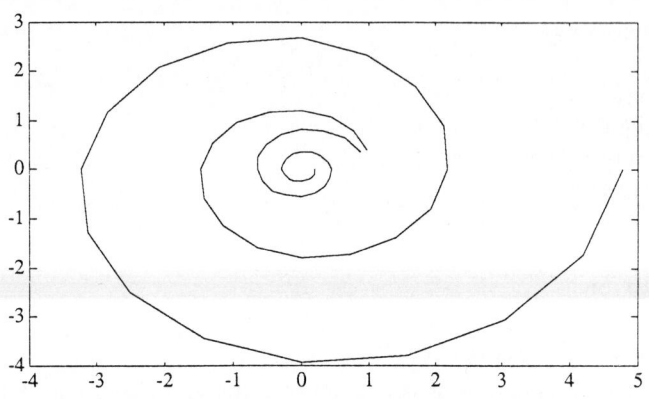

Figure 1.9: Powers of z

1.4 Roots of Quadratic Equations

You probably first encountered complex numbers when you studied values of z (called roots or zeros) for which the following equation is satisfied:

$$az^2 + bz + c = 0. \qquad 1.22$$

For $a \neq 0$ (as we will assume), this equation may be written as

$$z^2 + \frac{b}{a}z + \frac{c}{a} = 0. \qquad 1.23$$

Let's denote the second-degree polynomial on the left-hand side of this equation by $p(z)$:

$$p(z) = z^2 + \frac{b}{a}z + \frac{c}{a}. \qquad 1.24$$

This is called a *monic polynomial* because the coefficient of the highest-power term (z^2) is 1. When looking for solutions to the quadratic equation $z^2 + \frac{b}{a}z + $

$\frac{c}{a} = 0$, we are really looking for roots (or zeros) of the polynomial $p(z)$. The fundamental theorem of algebra says that there are two such roots. When we have found them, we may factor the polynomial $p(z)$ as follows:

$$p(z) = z^2 + \frac{b}{a}z + \frac{c}{a} = (z - z_1)(z - z_2). \qquad 1.25$$

In this equation, z_1 and z_2 are the roots we seek. The factored form $p(z) = (z - z_1)(z - z_2)$ shows clearly that $p(z_1) = p(z_2) = 0$, meaning that the quadratic equation $p(z) = 0$ is solved for $z = z_1$ and $z = z_2$. In the process of factoring the polynomial $p(z)$, we solve the quadratic equation and vice versa.

By equating the coefficients of z^2, z^1, and z^0 on the left- and right-hand sides of Equation 1.25, we find that the sum and the product of the roots z_1 and z_2 obey the equations

$$z_1 + z_2 = -\frac{b}{a} \qquad 1.26$$

$$z_1 z_2 = \frac{c}{a}.$$

You should always check your solutions with these equations.

Completing the Square. In order to solve the quadratic equation $z^2 + \frac{b}{a}z + \frac{c}{a} = 0$ (or, equivalently, to find the roots of the polynomial $z^2 + \frac{b}{a}z + \frac{c}{a}$), we "complete the square" on the left-hand side of Equation 1.23:

$$\left(z + \frac{b}{2a}\right)^2 - \left(\frac{b}{2a}\right)^2 + \frac{c}{a} = 0. \qquad 1.27$$

This equation may be rewritten as

$$\left(z + \frac{b}{2a}\right)^2 = \left(\frac{1}{2a}\right)^2 (b^2 - 4ac). \qquad 1.28$$

We may take the square root of each side to find the solutions

$$z_{1,2} = -\frac{b}{2a} \pm \frac{1}{2a}\sqrt{b^2 - 4ac}. \qquad 1.29$$

Problem 1.20 With the roots z_1 and z_2 defined in Equation 1.29, prove that $(z - z_1)(z - z_2)$ is, indeed, equal to the polynomial $z^2 + \frac{b}{a}z + \frac{c}{a}$. Check that $z_1 + z_2 = -\frac{b}{a}$ and $z_1 z_2 = \frac{c}{a}$. ∎

In the equation that defines the roots z_1 and z_2, the term $b^2 - 4ac$ is critical because it determines the nature of the solutions for z_1 and z_2. In fact, we may define three classes of solutions depending on $b^2 - 4ac$.

(i) Overdamped $(b^2 - 4ac > 0)$. In this case, the roots z_1 and z_2 are

$$z_{1,2} = -\frac{b}{2a} \pm \frac{1}{2a} \sqrt{b^2 - 4ac}. \qquad 1.30$$

These two roots are real, and they are located symmetrically about the point $-\frac{b}{2a}$. When $b = 0$, they are located symmetrically about 0 at the points $\pm \frac{1}{2a} \sqrt{-4ac}$. (In this case, $-4ac > 0$.) Typical solutions are illustrated in Figure 1.10.

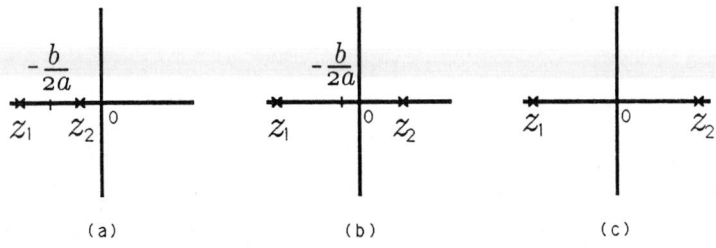

(a)   (b)   (c)

Figure 1.10: Typical Roots in the Overdamped Case; (a) $b/2a > 0, 4ac > 0$, (b) $b/2a > 0, 4ac < 0$, and (c) $b/2a = 0, 4ac < 0$

Problem 1.21 Compute and plot the roots of the following quadratic equations:

(a) $z^2 + 2z + \frac{1}{2} = 0$;
(b) $z^2 + 2a - \frac{1}{2} = 0$;
(c) $z^2 - \frac{1}{2} = 0$.

For each equation, check that $z_1 + z_2 = -\frac{b}{a}$ and $z_1 z_2 = \frac{c}{a}$. ∎

(ii) Critically Damped $(b^2 - 4ac = 0)$. In this case, the roots z_1 and z_2 are equal (we say they are repeated):

$$z_1 = z_2 = -\frac{b}{2a}. \qquad 1.31$$

These solutions are illustrated in Figure 1.11.

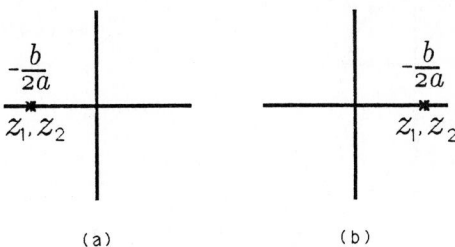

(a)   (b)

Figure 1.11: Roots in the Critically Damped Case; (a) $b/2a > 0$, and (b) $b/2a < 0$

Problem 1.22   Compute and plot the roots of the following quadratic equations:

(a) $z^2 + 2z + 1 = 0$;

(b) $z^2 - 2z + 1 = 0$;

(c) $z^2 = 0$.

For each equation, check that $z_1 + z_2 = -\frac{b}{a}$ and $z_1 z_2 = \frac{c}{a}$. ∎

(iii) Underdamped ($b^2 - 4ac < 0$). The underdamped case is, by far, the most fascinating case. When $b^2 - 4ac < 0$, then the square root in the solutions for z_1 and z_2 (Equation 1.29) produces an imaginary number. We may write $b^2 - 4ac$ as $-(4ac - b^2)$ and write $z_{1,2}$ as

$$
\begin{aligned}
z_{1,2} &= -\frac{b}{2a} \pm \frac{1}{2a}\sqrt{-(4ac - b^2)} \\
&= -\frac{b}{2a} \pm j\frac{1}{2a}\sqrt{4ac - b^2}.
\end{aligned}
$$

1.32

These complex roots are illustrated in Figure 1.12. Note that the roots are purely imaginary when $b = 0$, producing the result

$$
z_{1,2} = \pm j\sqrt{\frac{c}{a}}.
$$

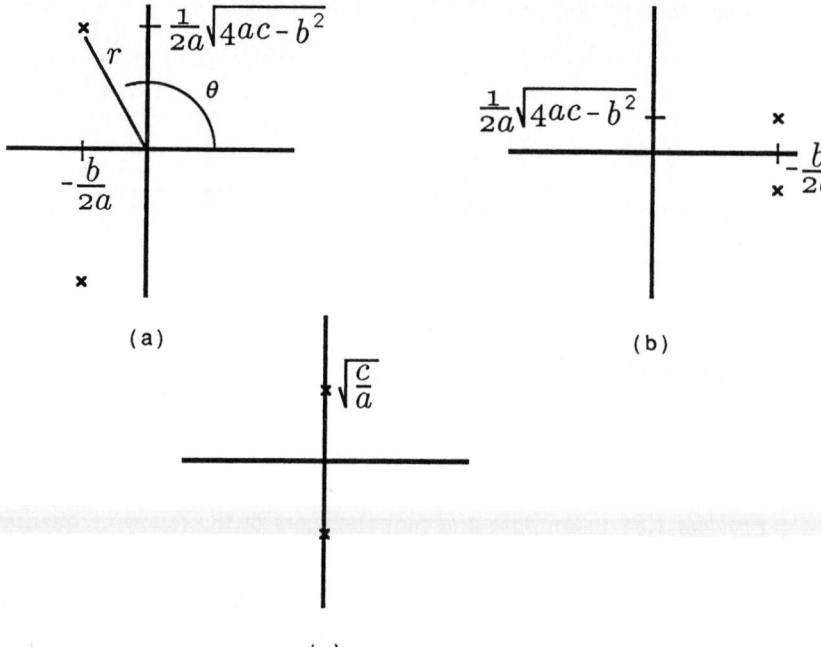

Figure 1.12: Typical Roots in the Underdamped Case; (a) $b/2a > 0$, (b) $b/2a < 0$, and (c) $b/2a = 0$

In this underdamped case, the roots z_1 and z_2 are complex conjugates:

$$z_2 = z_1^*. \tag{1.33}$$

Thus the polynomial $p(z) = z^2 + \frac{b}{a} z + \frac{c}{a} = (z - z_1)(z - z_2)$ also takes the form

$$p(z) = (z - z_1)(z - z_1^*)$$
$$= z^2 - 2\text{Re}[z_1]z + |z_1|^2. \tag{1.34}$$

$\text{Re}[z_1]$ and $|z_1|^2$ are related to the original coefficients of the polynomial as follows:

$$2\text{Re}[z_1] = -\frac{b}{a} \tag{1.35}$$

$$|z_1|^2 = \frac{c}{a}.$$

Always check these equations.

Let's explore these connections further by using the polar representations for z_1 and z_2:

$$z_{1,2} = re^{\pm j\theta}. \tag{1.36}$$

Then Equation 1.34 for the polynomial $p(z)$ may be written in the "standard form"

$$p(z) = (z - re^{j\theta})(z - re^{-j\theta})$$
$$= z^2 - 2r\cos\theta z + r^2. \tag{1.37}$$

Equation 1.35 is now

$$2r\cos\theta = -\frac{b}{a} \tag{1.38}$$

$$r^2 = \frac{c}{a}.$$

These equations may be used to locate $z_{1,2} = re^{\pm j\theta}$

$$r = \sqrt{\frac{c}{a}} \tag{1.39}$$

$$\theta = \pm\cos^{-1}\left(\frac{-b}{\sqrt{4ac}}\right).$$

Problem 1.23   Prove that $p(z)$ may be written as $p(z) = z^2 - 2r\cos\theta z + r^2$ in the underdamped case. ∎

Problem 1.24   Prove the relations in Equation 1.39. Outline a graphical procedure for locating $z_1 = re^{j\theta}$ and $z_2 = re^{-j\theta}$ from the polynomial $z^2 + \frac{b}{a}z + \frac{c}{a}$. ∎

Problem 1.25   Compute and plot the roots of the following quadratic equations:
  (a) $z^2 + 2z + 2 = 0$;
  (b) $z^2 - 2z + 2 = 0$;
  (c) $z^2 + 2 = 0$.
For each equation, check that $2\mathrm{Re}[z_{1,2}] = -\frac{b}{a}$ and $|z_{1,2}|^2 = \frac{c}{a}$. ∎

*1.5 Representing Complex Numbers in a Vector Space

So far we have coded the complex number $z = x + jy$ with the Cartesian pair (x, y) and with the polar pair $(r \angle \theta)$. We now show how the complex number z may be coded with a two-dimensional vector $\mathbf{z}$ and show how this new code may be used to gain insight about complex numbers.

Coding a Complex Number as a Vector. We code the complex number $z = x + jy$ with the two-dimensional vector $\mathbf{z} = \begin{bmatrix} x \\ y \end{bmatrix}$:

$$x + jy = z \quad \Longleftrightarrow \quad \mathbf{z} = \begin{bmatrix} x \\ y \end{bmatrix}. \tag{1.40}$$

We plot this vector as in Figure 1.13. We say that the vector $\mathbf{z}$ belongs to a "vector space." This means that vectors may be added and scaled according to the rules

$$\mathbf{z}_1 + \mathbf{z}_2 = \begin{bmatrix} x_1 + x_2 \\ y_1 + y_2 \end{bmatrix} \tag{1.41}$$

$$a\mathbf{z} = \begin{bmatrix} ax \\ ay \end{bmatrix}.$$

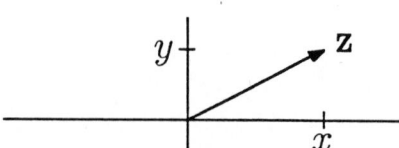

Figure 1.13: The Vector $\mathbf{z}$ Coding the Complex Number z

Furthermore, it means that an additive inverse $-\mathbf{z}$, an additive identity $\mathbf{0}$, and a multiplicative identity 1 all exist:

$$\mathbf{z} + (-\mathbf{z}) = \mathbf{0} \tag{1.42}$$

$$1\mathbf{z} = \mathbf{z}.$$

The vector $\mathbf{0}$ is $\mathbf{0} = \begin{bmatrix} 0 \\ 0 \end{bmatrix}$.

Problem 1.26 Prove that vector addition and scalar multiplication satisfy these properties of commutation, association, and distribution:

$$\mathbf{z}_1 + \mathbf{z}_2 = \mathbf{z}_2 + \mathbf{z}_1$$

$$(\mathbf{z}_1 + \mathbf{z}_2) + \mathbf{z}_3 = \mathbf{z}_1 + (\mathbf{z}_2 + \mathbf{z}_3)$$

$$a(b\mathbf{z}) = (ab)\mathbf{z}$$

$$a(\mathbf{z}_1 + \mathbf{z}_2) = a\mathbf{z}_1 + a\mathbf{z}_2. \quad \blacksquare$$

Inner Product and Norm. The *inner product* between two vectors $\mathbf{z}_1$ and $\mathbf{z}_2$ is defined to be the real number

$$(\mathbf{z}_1, \mathbf{z}_2) = x_1 x_2 + y_1 y_2. \qquad \qquad 1.43$$

We sometimes write this inner product as the vector product (more on this in Chapter 4)

$$(\mathbf{z}_1, \mathbf{z}_2) = \mathbf{z}_1^T \mathbf{z}_2$$

$$= [x_1 \ y_1] \begin{bmatrix} x_2 \\ y_2 \end{bmatrix} = (x_1 x_2 + y_1 y_2). \qquad \qquad 1.44$$

Problem 1.27 Prove $(\mathbf{z}_1, \mathbf{z}_2) = (\mathbf{z}_2, \mathbf{z}_1)$. $\quad \blacksquare$

When $\mathbf{z}_1 = \mathbf{z}_2 = \mathbf{z}$, then the inner product between $\mathbf{z}$ and itself is the *norm squared* of $\mathbf{z}$:

$$\| \mathbf{z} \|^2 = (\mathbf{z}, \mathbf{z}) = x^2 + y^2. \qquad \qquad 1.45$$

These properties of vectors seem abstract. However, as we now show, they may be used to develop a vector calculus for doing complex arithmetic.

A Vector Calculus for Complex Arithmetic. The addition of two complex numbers z_1 and z_2 corresponds to the addition of the vectors $\mathbf{z}_1$ and $\mathbf{z}_2$:

$$z_1 + z_2 \quad \Longleftrightarrow \quad \mathbf{z}_1 + \mathbf{z}_2 = \begin{bmatrix} x_1 + x_2 \\ y_1 + y_2 \end{bmatrix}. \qquad \qquad 1.46$$

The scalar multiplication of the complex number z_2 by the real number x_1 corresponds to scalar multiplication of the vector $\mathbf{z}_2$ by x_1:

$$x_1 z_2 \quad \Longleftrightarrow \quad x_1 \begin{bmatrix} x_2 \\ y_2 \end{bmatrix} = \begin{bmatrix} x_1 x_2 \\ x_1 y_2 \end{bmatrix}. \qquad 1.47$$

Similarly, the multiplication of the complex number z_2 by the real number y_1 is

$$y_1 z_2 \quad \leftrightarrow \quad y_1 \begin{bmatrix} x_2 \\ y_2 \end{bmatrix} = \begin{bmatrix} y_1 x_2 \\ y_1 y_2 \end{bmatrix}. \qquad 1.48$$

The complex product $z_1 z_2 = (x_1 + j y_1) z_2$ is therefore represented as

$$z_1 z_2 \quad \leftrightarrow \quad \begin{bmatrix} x_1 x_2 - y_1 y_2 \\ x_1 y_2 + y_1 x_2 \end{bmatrix}. \qquad 1.49$$

This representation may be written as the inner product

$$z_1 z_2 = z_2 z_1 \quad \leftrightarrow \quad \begin{bmatrix} (\mathbf{v}, \mathbf{z}_1) \\ (\mathbf{w}, \mathbf{z}_1) \end{bmatrix} \qquad 1.50$$

where $\mathbf{v}$ and $\mathbf{w}$ are the vectors $\mathbf{v} = \begin{bmatrix} x_2 \\ -y_2 \end{bmatrix}$ and $\mathbf{w} = \begin{bmatrix} y_2 \\ x_2 \end{bmatrix}$. By defining the matrix

$$\begin{bmatrix} x_2 & -y_2 \\ y_2 & x_2 \end{bmatrix}, \qquad 1.51$$

we can represent the complex product $z_1 z_2$ as a matrix-vector multiply (more on this in Chapter 4):

$$z_1 z_2 = z_2 z_1 \quad \leftrightarrow \quad \begin{bmatrix} x_2 & -y_2 \\ y_2 & x_2 \end{bmatrix} \begin{bmatrix} x_1 \\ y_1 \end{bmatrix}. \qquad 1.52$$

With this representation, we can represent rotation as

$$z e^{j\theta} = e^{j\theta} z \quad \leftrightarrow \quad \begin{bmatrix} \cos\theta & -\sin\theta \\ \sin\theta & \cos\theta \end{bmatrix} \begin{bmatrix} x_1 \\ x_2 \end{bmatrix}. \qquad 1.53$$

We call the matrix $\begin{bmatrix} \cos\theta & -\sin\theta \\ \sin\theta & \cos\theta \end{bmatrix}$ a "rotation matrix."

Problem 1.28 Call $\mathbf{R}(\theta)$ the rotation matrix:

$$\mathbf{R}(\theta) = \begin{bmatrix} \cos\theta & -\sin\theta \\ \sin\theta & \cos\theta \end{bmatrix}.$$

Show that $\mathbf{R}(-\theta)$ rotates by $(-\theta)$. What can you say about $\mathbf{R}(-\theta)\mathbf{w}$ when $\mathbf{w} = \mathbf{R}(\theta)\mathbf{z}$? ∎

Problem 1.29 Represent the complex conjugate of z as

$$z^* \quad \leftrightarrow \quad \begin{bmatrix} a & b \\ c & d \end{bmatrix} \begin{bmatrix} x \\ y \end{bmatrix}$$

and find the elements a, b, c, and d of the matrix. ∎

Inner Product and Polar Representation. From the norm of a vector, we derive a formula for the magnitude of z in the polar representation $z = re^{j\theta}$:

$$r = (x^2 + y^2)^{1/2} = \|\mathbf{z}\| = (\mathbf{z}, \mathbf{z})^{1/2}. \qquad 1.54$$

If we define the coordinate vectors $\mathbf{e}_1 = \begin{bmatrix} 1 \\ 0 \end{bmatrix}$ and $\mathbf{e}_2 = \begin{bmatrix} 0 \\ 1 \end{bmatrix}$, then we can represent the vector $\mathbf{z}$ as

$$\mathbf{z} = (\mathbf{z}, \mathbf{e}_1)\mathbf{e}_1 + (\mathbf{z}, \mathbf{e}_2)\mathbf{e}_2. \qquad 1.55$$

See Figure 1.14. From the figure it is clear that the cosine and sine of the angle θ are

$$\cos\theta = \frac{(\mathbf{z}, \mathbf{e}_1)}{\|\mathbf{z}\|}; \qquad \sin\theta = \frac{(\mathbf{z}, \mathbf{e}_2)}{\|\mathbf{z}\|}. \qquad 1.56$$

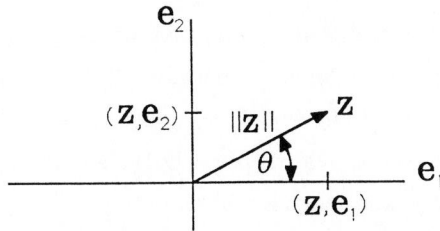

Figure 1.14: Representation of $\mathbf{z}$ in its Natural Basis

This gives us another representation for any vector $\mathbf{z}$:

$$\mathbf{z} = \|\mathbf{z}\| \cos\theta \mathbf{e}_1 + \|\mathbf{z}\| \sin\theta \mathbf{e}_2. \qquad 1.57$$

The inner product between two vectors $\mathbf{z}_1$ and $\mathbf{z}_2$ is now

$$\begin{aligned}
(\mathbf{z}_1, \mathbf{z}_2) &= \left[(\mathbf{z}_1, \mathbf{e}_1)\mathbf{e}_1^T \quad (\mathbf{z}_1, \mathbf{e}_2)\mathbf{e}_2^T\right]\begin{bmatrix}(\mathbf{z}_2, \mathbf{e}_1)\mathbf{e}_1 \\ (\mathbf{z}_2, \mathbf{e}_2)\mathbf{e}_2\end{bmatrix} \\
&= (\mathbf{z}_1, \mathbf{e}_1)(\mathbf{z}_2, \mathbf{e}_1) + (\mathbf{z}_1, \mathbf{e}_2)(\mathbf{z}_2, \mathbf{e}_2) \qquad\qquad 1.58 \\
&= \|\mathbf{z}_1\| \cos\theta_1 \|\mathbf{z}_2\| \cos\theta_2 + \|\mathbf{z}_1\| \sin\theta_1 \|\mathbf{z}_2\| \sin\theta_2.
\end{aligned}$$

It follows that $\cos(\theta_2 - \theta_1) = \cos\theta_2 \cos\theta_1 + \sin\theta_1 \sin\theta_2$ may be written as

$$\cos(\theta_2 - \theta_1) = \frac{(\mathbf{z}_1, \mathbf{z}_2)}{\|\mathbf{z}_1\| \, \|\mathbf{z}_2\|}. \qquad 1.59$$

This formula shows that the cosine of the angle between two vectors $\mathbf{z}_1$ and $\mathbf{z}_2$, which is, of course, the cosine of the angle of $z_2 z_1^*$, is the ratio of the inner product to the norms.

Problem 1.30 Prove the Schwarz and triangle inequalities and interpret them:

$$\text{(Schwarz)} \qquad (\mathbf{z}_1, \mathbf{z}_2)^2 \leq \|\mathbf{z}_1\|^2 \|\mathbf{z}_2\|^2$$

$$\text{(triangle)} \qquad \|\mathbf{z}_1 - \mathbf{z}_2\| \leq \|\mathbf{z}_1 - \mathbf{z}_3\| + \|\mathbf{z}_2 - \mathbf{z}_3\|. \quad \blacksquare$$

*1.6 An Electric Field Computation

We have established that vectors may be used to code complex numbers. Conversely, complex numbers may be used to code or represent the orthogonal components of any two-dimensional vector. This makes them invaluable in electromagnetic field theory, where they are used to represent the components of electric and magnetic fields.

The basic problem in electromagnetic field theory is to determine the electric or magnetic field that is generated by a static or dynamic distribution of charge. The key idea is to isolate an infinitesimal charge, determine the field set up by this charge, and then to sum the fields contributed by all such infinitesimal charges. This idea is illustrated in Figure 1.15, where the charge λ, uniformly distributed over a line segment of length dx at point $-x$, produces a field $dE(x)$ at the test point $(0, h)$. The field $dE(x)$ is a "vector" field (as opposed to a "scalar" field), with components $E_1(x)$ and $E_2(x)$. The intensity or field strength of the field $dE(x)$ is

$$|dE(x)| = \frac{\lambda\,dx}{4\pi\epsilon_0(h^2 + x^2)}. \qquad 1.60$$

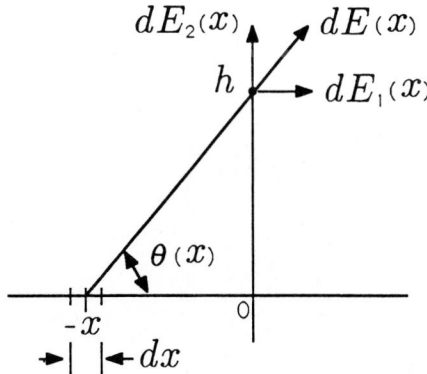

Figure 1.15: Infinitesimal Charge $\lambda\,dx$ Producing Field $dE(x)$

But the field strength is directed at angle $\theta(x)$, as illustrated in Figure 1.15. The field $dE(x)$ is real with components $dE_1(x)$ and $dE_2(x)$, but we code it as a complex field. We say that the "complex" field at test point $(0, h)$ is

$$dE(x) = \frac{\lambda\,dx}{4\pi\epsilon_0(h^2 + x^2)}\,e^{j\theta(x)} \qquad 1.61$$

with components $dE_1(x)$ and $dE_2(x)$. That is,

$$dE(x) = dE_1(x) + j\,dE_2(x) \qquad 1.62$$

$$dE_1(x) = \frac{\lambda\,dx}{4\pi\epsilon_0(h^2 + x^2)}\,\cos\theta(x)$$

$$dE_2(x) = \frac{\lambda\,dx}{4\pi\epsilon_0(h^2 + x^2)}\sin\theta(x).$$

For charge uniformally distributed with density λ along the x-axis, the total field at the test point $(0, h)$ is obtained by integrating dE:

$$E = \int_{-\infty}^{\infty} dE(x) = \int_{-\infty}^{\infty} \frac{\lambda}{4\pi\epsilon_0(h^2 + x^2)} \left[\cos\theta(x) + j\sin\theta(x)\right] dx. \qquad 1.63$$

The functions $\cos\theta(x)$ and $\sin\theta(x)$ are

$$\cos\theta(x) = \frac{x}{(x^2 + h^2)^{1/2}}\;; \qquad \sin\theta(x) = \frac{h}{(x^2 + h^2)^{1/2}}. \qquad 1.64$$

We leave it as a problem to show that the real component E_1 of the field is zero. The imaginary component E_2 is

$$
\begin{aligned}
E = jE_2 &= j \int_{-\infty}^{\infty} \frac{\lambda h}{4\pi\epsilon_0}\frac{dx}{(x^2 + h^2)^{3/2}} \\
&= j\,\frac{\lambda h}{4\pi\epsilon_0}\frac{x}{h^2(x^2 + h^2)^{1/2}}\Bigg|_{-\infty}^{\infty} \qquad 1.65 \\
&= j\,\frac{\lambda h}{4\pi\epsilon_0}\left[\frac{1}{h^2} + \frac{1}{h^2}\right] = j\,\frac{\lambda}{2\pi\epsilon_0 h}
\end{aligned}
$$

$$E_2 = \frac{\lambda}{2\pi\epsilon_0 h}.$$

We emphasize that the field at $(0, h)$ is a *real* field. Our imaginary answer simply says that the real field is *oriented* in the vertical direction because we have used the imaginary part of the complex field to code the vertical component of the real field.

Problem 1.31   Show that the horizontal component of the field E is zero. Interpret this finding physically. ∎

From the symmetry of this problem, we conclude that the field around the infinitely long wire of Figure 1.15 is radially symmetric. So, in polar coordinates, we could say

$$E(r, \theta) = \frac{\lambda}{2\pi\epsilon_0 r}, \qquad 1.66$$

which is independent of θ. If we integrated the field along a radial line perpendicular to the wire, we would measure the voltage difference

$$V(r_1) - V(r_0) = \int_{r_0}^{r_1} \frac{\lambda}{2\pi\epsilon_0 r} \, dr = \frac{\lambda}{2\pi\epsilon_0} [\log r_1 - \log r_0]. \qquad 1.67$$

An electric field has units of volts/meter, a charge density λ has units of coulombs/meter, and ϵ_0 has units of coulombs/volt-meter; voltage has units of volts (of course).

1.7 Numerical Experiment (Quadratic Roots)

There is a version of the quadratic equation that will arise over and over again in your study of electrical and mechanical systems:

$$s^2 + 2\xi\omega_0 s + \omega_0^2 = 0. \qquad 1.68$$

For reasons that can only become clear as you continue your study of engineering, the parameter ω_0 is called a *resonant frequency*, and the parameter $\xi \geq 0$ is called a *damping factor*. In this experiment, you will begin by

(1) finding the "underdamped" range of values $\xi \geq 0$ for which the roots s_1 and s_2 are complex;

(2) finding the "critically damped" value of $\xi \geq 0$ that makes the roots s_1 and s_2 equal; and

(3) finding the "overdamped" range of values $\xi \geq 0$ for which s_1 and s_2 are real.

For each of these ranges,

(4) find the analytical solution for $s_{1,2}$ as a function of ω_0 and ξ; write your solutions in Cartesian and polar forms and present your results as

$$s_{1,2} = \begin{cases} & , \quad 0 \leq \xi \leq \xi_c \\ & , \quad \xi = \xi_c \\ & , \quad \xi \geq \xi_c \end{cases}$$

where ξ_c is the critically damped value of ξ. Write a MATLAB program that computes and plots $s_{1,2}$ for ω_0 fixed at $\omega_0 = 1$ and ξ variable between 0.0 and 2.0 in steps of 0.1. Interpret all of your findings.

Now organize the coefficients of the polynomial $s^2 + 2\xi s + 1$ into the array $[1\ 2\xi\ 1]$. Imbed the MATLAB instructions

```
r=roots([1 2*e 1]);
plot(real(r(1)),imag(r(1)),'o')
plot(real(r(2)),imag(r(2)),'o')
```

in a `for` loop to compute and plot the roots of $s^2 + 2\xi s + 1$ as ξ ranges from 0.0 to 2.0. Note that r is a 1×2 array of complex numbers. You should observe Figure 1.16. We call this "half circle and line" the locus of roots for the quadratic equation or the "root locus" in shorthand.

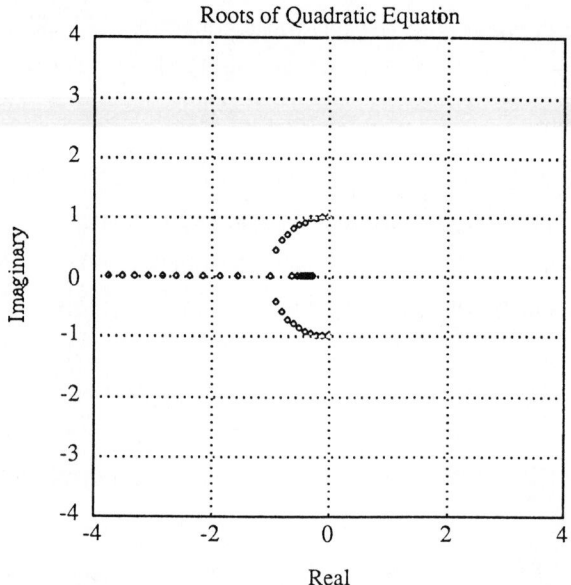

Figure 1.16: Roots of Quadratic Equation

2

The Functions
e^x and $e^{j\theta}$

Notes to Teachers and Students:

It is essential to write out, term-by-term, every sequence and sum in this chapter. This demystifies the seemingly mysterious notation. The example on compound interest shows the value of limiting arguments in everyday life and gives e^x some real meaning. The function $e^{j\theta}$, covered in Sections 2.3 and 2.7, must be understood by all students before proceeding to Chapter 3. The Euler and De Moivre identities provide every tool that students need to derive trigonometric formulas. The properties of roots of unity are invaluable for the study of phasors in Chapter 3.

The MATLAB programs in this chapter are used to illustrate sequences and series and to explore approximations to $\sin\theta$ and $\cos\theta$. The numerical experiment in Section 2.7 illustrates, geometrically and algebraically, how approximations to $e^{j\theta}$ converge.

Section 2.6, "Second-Order Differential and Difference Equations," is a little demanding for freshmen, but we give it a once-over-lightly to illustrate the power of quadratic equations and the functions e^x and $e^{j\theta}$. This section also gives a sneak preview of more advanced courses in circuits and systems.

2.1 Introduction

It is probably not too strong a statement to say that the function e^x is the most important function in engineering and applied science. In this chapter we study the function e^x and extend its definition to the function $e^{j\theta}$. This study clarifies our definition of $e^{j\theta}$ from Chapter 1 and leads us to an investigation of sequences and series. We use the function $e^{j\theta}$ to derive the Euler and De Moivre identities and to produce a number of important trigonometric identities. We define the complex roots of unity and study their partial sums. The results of this chapter will be used in Chapter 3 when we study the phasor representation of sinusoidal signals.

2.2 The Function e^x

Many of you know the number e as the base of the natural logarithm, which has the value $2.718281828459045\ldots$. What you may not know is that this number is actually defined as the limit of a sequence of approximating numbers. That is,

$$e = \lim_{n \to \infty} f_n \qquad\qquad 2.1$$

$$f_n = \left(1 + \frac{1}{n}\right)^n, \qquad n = 1, 2, \ldots .$$

This means, simply, that the sequence of numbers $(1+1)^1$, $\left(1+\frac{1}{2}\right)^2$, $\left(1+\frac{1}{3}\right)^3$, $\ldots$, gets arbitrarily close to $2.718281828459045\ldots$. But why should such a sequence of numbers be so important? In the next several paragraphs we answer this question.

Problem 2.1 (MATLAB) Write a MATLAB program to evaluate the expression $f_n = \left(1 + \frac{1}{n}\right)^n$ for $n = 1, 2, 4, 8, 16, 32, 64$ to show that $f_n \approx e$ for large n. ∎

Derivatives and the Number e. The number $f_n = \left(1 + \frac{1}{n}\right)^n$ arises in the study of derivatives in the following way. Consider the function

$$f(x) = a^x, \qquad a > 1 \qquad\qquad 2.2$$

and ask yourself when the derivative of $f(x)$ equals $f(x)$. The function $f(x)$ is plotted in Figure 2.1 for $a > 1$. The slope of the function at point x is

$$\frac{df(x)}{dx} = \lim_{\Delta x \to 0} \frac{a^{x+\Delta x} - a^x}{\Delta x}$$
$$= a^x \lim_{\Delta x \to 0} \frac{a^{\Delta x} - 1}{\Delta x}. \qquad\qquad 2.3$$

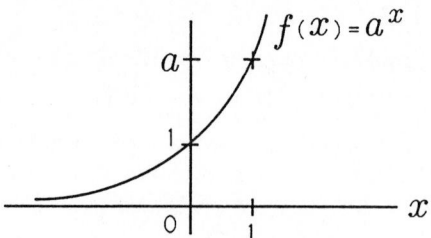

Figure 2.1: The Function $f(x) = a^x$

If there is a special value for a such that

$$\lim_{\Delta x \to 0} \frac{a^{\Delta x} - 1}{\Delta x} = 1,$$

2.4

then $\frac{d}{dx} f(x)$ would equal $f(x)$. We call this value of a the special (or exceptional) number e and write

$$f(x) = e^x$$

2.5

$$\frac{d}{dx} f(x) = e^x.$$

The number e would then be $e = f(1)$. Let's write our condition that $\frac{a^{\Delta x} - 1}{\Delta x}$ converges to 1 as

$$e^{\Delta x} - 1 \cong \Delta x, \qquad \Delta x \text{ small}$$

2.6

or as

$$e \cong (1 + \Delta x)^{1/\Delta x}.$$

2.7

Our definition of $e = \lim_{n \to \infty} \left(1 + \frac{1}{n}\right)^{1/n}$ amounts to defining $\Delta x = \frac{1}{n}$ and allowing $n \to \infty$ in order to make $\Delta x \to 0$. With this definition for e, it is clear that the function e^x is defined to be $(e)^x$:

$$e^x = \lim_{\Delta x \to 0} (1 + \Delta x)^{x/\Delta x}.$$

2.8

By letting $\Delta x = \frac{x}{n}$, we can write this definition in the more familiar form

$$e^x = \lim_{n \to \infty} \left(1 + \frac{x}{n}\right)^n.$$

2.9

This is our fundamental definition for the function e^x. When evaluated at $x = 1$, it produces the definition of e given in Equation 2.1.

The derivative of e^x is, of course,

$$\frac{d}{dx} e^x = \lim_{n \to \infty} n \left(1 + \frac{x}{n}\right)^{n-1} \frac{1}{n} = e^x. \qquad 2.10$$

This means that Taylor's theorem[1] may be used to find another characterization for e^x:

$$e^x = \sum_{n=0}^{\infty} \left[\frac{d^n}{dx^n} e^x\right]_{x=0} \frac{1}{n!} x^n$$
$$= \sum_{n=0}^{\infty} \frac{x^n}{n!}. \qquad 2.11$$

When this series expansion for e^x is evaluated at $x = 1$, it produces the following series for e:

$$e = \sum_{n=0}^{\infty} \frac{1}{n!}. \qquad 2.12$$

In this formula, $n!$ is the product $n(n-1)(n-2)\cdots(2)1$. Read $n!$ as "n factorial."

Problem 2.2 (MATLAB) Write a MATLAB program to evaluate the sum

$$S_N = \sum_{n=0}^{N} \frac{1}{n!}$$

for $N = 1, 2, 4, 8, 16, 32, 64$ to show that $S_N \cong e$ for large N. Compare S_{64} with f_{64} from Problem 2.1. Which approximation do you prefer? ∎

Compound Interest and the Function e^x. There is an example from your everyday life that shows even more dramatically how the function e^x arises. Suppose you invest V_0 dollars in a savings account that offers $100x\%$ annual interest. (When $x = 0.01$, this is 1%; when $x = 0.10$, this is 10% interest.) If interest is compounded only once per year, you have the simple

---

[1] Taylor's theorem says that a function may be completely characterized by all of its derivatives (provided they all exist). See Appendix C.

interest formula for V_1, the value of your savings account after 1 compound (in this case, 1 year):

$$V_1 = (1 + x)V_0. \qquad 2.13$$

This result is illustrated in the block diagram of Figure 2.2(a). In this diagram, your input fortune V_0 is processed by the "interest block" to produce your output fortune V_1. If interest is compounded monthly, then the annual interest is divided into 12 equal parts and applied 12 times. The compounding formula for V_{12}, the value of your savings after 12 compounds (also 1 year) is

$$V_{12} = \left(1 + \frac{x}{12}\right)^{12} V_0. \qquad 2.14$$

This result is illustrated in Figure 2.2(b). Can you read the block diagram? The general formula for the value of an account that is compounded n times per year is

$$V_n = \left(1 + \frac{x}{n}\right)^n V_0. \qquad 2.15$$

V_n is the value of your account after n compounds in a year, when the annual interest rate is $100x\%$.

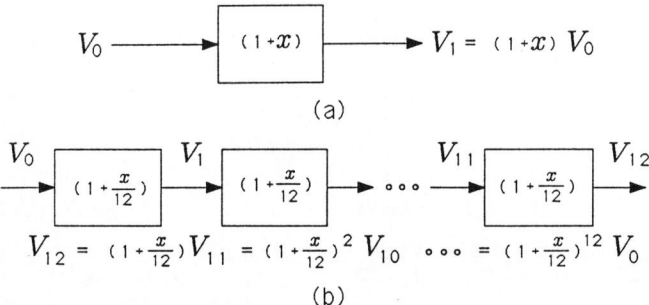

Figure 2.2: Block Diagram for Interest Computations; (a) Simple Annual Interest, and (b) Monthly Compounding

Problem 2.3   Verify in Equation 2.15 that a recursion is at work that terminates at V_n. That is, show that $V_{i+1} = \left(1 + \frac{x}{n}\right)V_i$ for $i = 0, 1, \ldots, n-1$ produces the result $V_n = \left(1 + \frac{x}{n}\right)^n V_0$. ∎

Bankers have discovered the (apparent) appeal of infinite, or continuous, compounding:

$$V_\infty = \lim_{n\to\infty} \left(1 + \frac{x}{n}\right)^n V_0. \qquad 2.16$$

We know that this is just

$$V_\infty = e^x V_0. \qquad 2.17$$

So, when deciding between $100x_1\%$ interest compounded daily and $100x_2\%$ interest compounded continuously, we need only compare

$$\left(1 + \frac{x_1}{365}\right)^{365} \quad \text{versus} \quad e^{x_2}. \qquad 2.18$$

We suggest that daily compounding is about as good as continuous compounding. What do you think? How about monthly compounding?

Problem 2.4 (MATLAB) Write a MATLAB program to compute and plot simple interest, monthly interest, daily interest, and continuous interest versus interest rate $100x$. Use the curves to develop a strategy for saving money. ∎

2.3 The Function $e^{j\theta}$ and the Unit Circle

Let's try to extend our definitions of the function e^x to the argument $x = j\theta$. Then $e^{j\theta}$ is the function

$$e^{j\theta} = \lim_{n\to\infty} \left(1 + j\frac{\theta}{n}\right)^n. \qquad 2.19$$

The complex number $1 + j\frac{\theta}{n}$ is illustrated in Figure 2.3. The radius to the point $1 + j\frac{\theta}{n}$ is $r = \left(1 + \frac{\theta^2}{n^2}\right)^{1/2}$, and the angle is $\phi = \tan^{-1}\frac{\theta}{n}$. This means that the nth power of $1 + j\frac{\theta}{n}$ has radius $r^n = \left(1 + \frac{\theta^2}{n^2}\right)^{n/2}$ and angle $n\phi = n\tan^{-1}\frac{\theta}{n}$. (Recall our study of powers of z.) Therefore the complex number $\left(1 + j\frac{\theta}{n}\right)^n$ may be written as

$$\left(1 + j\frac{\theta}{n}\right)^n = \left(1 + \frac{\theta^2}{n^2}\right)^{n/2} \left[\cos\left(n\tan^{-1}\frac{\theta}{n}\right) + j\sin\left(n\tan^{-1}\frac{\theta}{n}\right)\right]. \qquad 2.20$$

For n large, $\left(1 + \frac{\theta^2}{n^2}\right)^{n/2} \cong 1$, and $n\tan^{-1}\frac{\theta}{n} \cong n\frac{\theta}{n} = \theta$. Therefore $\left(1 + j\frac{\theta}{n}\right)^n$ is approximately

$$\left(1 + j\frac{\theta}{n}\right)^n = 1(\cos\theta + j\sin\theta). \qquad 2.21$$

This finding is consistent with our previous definition of $e^{j\theta}$!

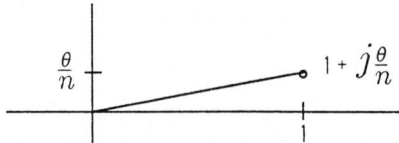

Figure 2.3: The Complex Number $1 + j\,\dfrac{\theta}{n}$

The series expansion for $e^{j\theta}$ is obtained by evaluating Taylor's formula at $x = j\theta$:

$$e^{j\theta} = \sum_{n=0}^{\infty} \frac{1}{n!} (j\theta)^n.$$

2.22

When this series expansion for $e^{j\theta}$ is written out, we have the formula

$$
\begin{aligned}
e^{j\theta} &= \sum_{n=0}^{\infty} \frac{1}{(2n)!} (j\theta)^{2n} + \sum_{n=0}^{\infty} \frac{1}{(2n+1)!} (j\theta)^{2n+1} \\
&= \sum_{n=0}^{\infty} \frac{(-1)^n}{(2n)!} \theta^{2n} + j \sum_{n=0}^{\infty} \frac{(-1)^n}{(2n+1)!} \theta^{2n+1}.
\end{aligned}
$$

2.23

It is now clear that $\cos\theta$ and $\sin\theta$ have the series expansions

$$\cos\theta = \sum_{n=0}^{\infty} \frac{(-1)^n}{(2n)!} \theta^{2n}$$

2.24

$$\sin\theta = \sum_{n=0}^{\infty} \frac{(-1)^n}{(2n+1)!} \theta^{2n+1}.$$

When these infinite sums are truncated at $N - 1$, then we say that we have N-term approximations for $\cos\theta$ and $\sin\theta$:

$$\cos\theta \cong \sum_{n=0}^{N-1} \frac{(-1)^n}{(2n)!} \theta^{2n}$$

2.25

$$\sin\theta \cong \sum_{n=0}^{N-1} \frac{(-1)^n}{(2n+1)!} \theta^{2n+1}.$$

The ten-term approximations to $\cos\theta$ and $\sin\theta$ are plotted over exact expressions for $\cos\theta$ and $\sin\theta$ in Figure 2.4. The approximations are very good over one period ($0 \le \theta \le 2\pi$), but they diverge outside this interval. For more accurate approximations over a larger range of θ's, we would need to use more terms. Or, better yet, we could use the fact that $\cos\theta$ and $\sin\theta$ are periodic in θ. Then we could subtract as many multiples of 2π as we needed from θ to bring the result into the range $[0, 2\pi]$ and use the ten-term approximations on this new variable. The new variable is called "θ-modulo 2π."

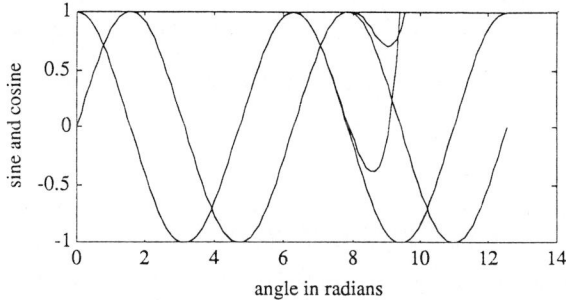

Figure 2.4: Ten-Term Approximations to $\cos\theta$ and $\sin\theta$

Problem 2.5 Write out the first several terms in the series expansions for $\cos\theta$ and $\sin\theta$. ∎

Demo 2.1 (MATLAB). Create a MATLAB file containing the following demo MATLAB program that computes and plots two cycles of $\cos\theta$ and $\sin\theta$ versus θ. You should observe Figure 2.5. Note that two cycles take in $2(2\pi)$ radians, which is approximately 12 radians.

```
clg;
j = sqrt(-1);
theta = 0:2*pi/50:4*pi;
s = sin(theta);
c = cos(theta);
plot(theta,s);
xlabel('theta in radians');
```

```
ylabel('sine and cosine');
hold on
plot(theta,c);
hold off  □
```

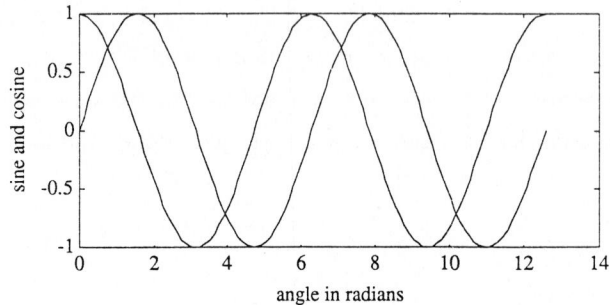

Figure 2.5: The Functions $\cos\theta$ and $\sin\theta$

Problem 2.6 (MATLAB) Write a MATLAB program to compute and plot the ten-term approximations to $\cos\theta$ and $\sin\theta$ for θ running from 0 to $2(2\pi)$ in steps of $2\pi/50$. Compute and overplot exact expressions for $\cos\theta$ and $\sin\theta$. You should observe a result like Figure 2.4. ∎

The Unit Circle. The *unit circle* is defined to be the set of all complex numbers z whose magnitudes are 1. This means that all the numbers on the unit circle may be written as $z = e^{j\theta}$. We say that the unit circle consists of all numbers generated by the function $z = e^{j\theta}$ as θ varies from 0 to 2π. See Figure 2.6.

A Fundamental Symmetry. Let's consider the two complex numbers z_1 and $\frac{1}{z_1^*}$, illustrated in Figure 2.6. We call $\frac{1}{z_1^*}$ the "reflection of z through the unit circle" (and vice versa). Note that $z_1 = r_1 e^{j\theta_1}$ and $\frac{1}{z_1^*} = \frac{1}{r_1} e^{j\theta_1}$. The complex numbers $z_1 - e^{j\theta}$ and $\frac{1}{z_1^*} - e^{j\theta}$ are illustrated in Figure 2.6. The magnitude squared of each is

$$|z_1 - e^{j\theta}|^2 = (z_1 - e^{j\theta})(z_1^* - e^{-j\theta}) \qquad 2.26$$

$$\left|\frac{1}{z_1^*} - e^{j\theta}\right|^2 = \left(\frac{1}{z_1^*} - e^{j\theta}\right)\left(\frac{1}{z_1} - e^{-j\theta}\right). \qquad 2.27$$

The ratio of these magnitudes squared is

$$\beta^2 = \frac{(z_1 - e^{j\theta})(z_1^* - e^{-j\theta})}{\left(\frac{1}{z_1^*} - e^{j\theta}\right)\left(\frac{1}{z_1} - e^{-j\theta}\right)}. \qquad 2.28$$

This ratio may be manipulated to show that it is independent of θ, meaning that the points z_1 and $\frac{1}{z_1^*}$ maintain a constant relative distance from every point on the unit circle:

$$\beta^2 = \frac{e^{j\theta}(e^{-j\theta}z_1 - 1)(z_1^* e^{j\theta} - 1)e^{-j\theta}}{\frac{1}{z_1^*}(1 - e^{j\theta}z_1^*)(1 - z_1 e^{-j\theta})\frac{1}{z_1}}$$
$$= |z_1|^2, \qquad \text{independent of } \theta! \qquad 2.29$$

This result will be of paramount importance to you when you study digital filtering, antenna design, and communication theory.

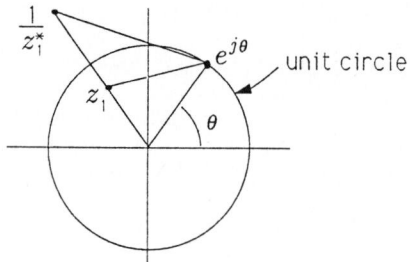

Figure 2.6: The Unit Circle

Problem 2.7   Write the complex number $z - e^{j\theta}$ as $re^{j\phi}$. What are r and ϕ? ∎

2.4 The Euler and De Moivre Identities

The Euler and De Moivre identities are the fundamental identities for deriving trigonometric formulas. From the identity $e^{j\theta} = \cos\theta + j\sin\theta$ and the conjugate identity $e^{-j\theta} = (e^{j\theta})^* = \cos\theta - j\sin\theta$, we have the Euler identities for $\cos\theta$ and $\sin\theta$:

$$\cos\theta = \frac{e^{j\theta} + e^{-j\theta}}{2} \qquad\qquad 2.30$$

$$\sin\theta = \frac{e^{j\theta} - e^{-j\theta}}{2j}.$$

These identities are illustrated in Figure 2.7.

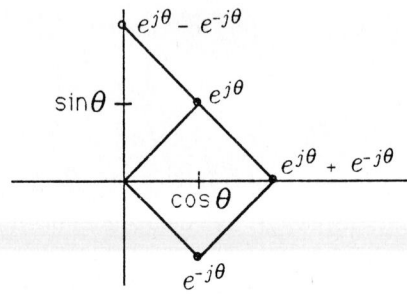

Figure 2.7: Euler's Identities

The identity $e^{j\theta} = \cos\theta + j\sin\theta$ also produces the De Moivre identity:

$$(\cos\theta + j\sin\theta)^n = (e^{j\theta})^n = e^{jn\theta}$$

$$= \cos n\theta + j\sin n\theta. \qquad\qquad 2.31$$

When the left-hand side of this equation is expanded with the binomial expansion, we obtain the identity

$$\sum_{k=0}^{n}\binom{n}{k}(\cos\theta)^{n-k}(j\sin\theta)^k = \cos n\theta + j\sin n\theta. \qquad\qquad 2.32$$

Binomial Coefficients and Pascal's Triangle. The binomial coefficients $\binom{n}{k}$ in Equation 2.32 are shorthand for the number

$$\binom{n}{k} = \frac{n!}{(n-k)!k!}, \qquad k = 0, 1, \ldots, n.$$

This number gives the coefficient of $x^{n-k}y^k$ in the expansion of $(x+y)^n$. How do we know that there are $\binom{n}{k}$ terms of the form $x^{n-k}y^k$? One way to answer this question is to use Pascal's triangle, illustrated in Figure 2.8. Each node on Pascal's triangle shows the number of routes that terminate at that node. This number is always the sum of the number of routes that terminate at the nodes just above the node in question. If we think of a left-hand path as an occurrence of an x and a right-hand path as an occurrence of a y, then we see that Pascal's triangle keeps track of the number of occurrences of $x^{n-k}y^k$.

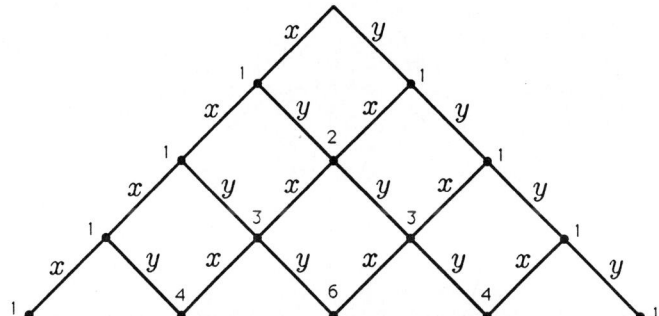

Figure 2.8: Pascal's Triangle and the Binomial Coefficients

Problem 2.8   Prove $\binom{n}{k} = \binom{n}{n-k}$. ∎

Problem 2.9   Find an identity for $\binom{n-1}{k-1} + \binom{n-1}{k}$. ∎

Problem 2.10   Find "half-angle" formulas for $\cos 2\theta$ and $\sin 2\theta$. ∎

Problem 2.11   Show that
  (a) $\cos 3\theta = \cos^2 \theta - 3\cos\theta \sin^2 \theta$;
  (b) $\sin 3\theta = 3\cos^2 \theta \sin\theta - \sin^3 \theta$. ∎

Problem 2.12   Use $e^{j(\theta_1 + \theta_2)} = e^{j\theta_1} e^{j\theta_2} = (\cos\theta_1 + j\sin\theta_1)(\cos\theta_2 + j\sin\theta_2)$ to prove
  (a) $\cos(\theta_1 + \theta_2) = \cos\theta_1 \cos\theta_2 - \sin\theta_1 \sin\theta_2$;
  (b) $\sin(\theta_1 + \theta_2) = \sin\theta_1 \cos\theta_2 + \sin\theta_2 \cos\theta_1$. ∎

2.5 Roots of Unity and Related Topics

The complex number $z = e^{j2\pi/N}$ is illustrated in Figure 2.9. It lies on the unit circle at angle $\theta = 2\pi/N$. When this number is raised to the nth power, the result is $z^n = e^{j2\pi n/N}$. This number is also illustrated in Figure 2.9. When one of the complex numbers $e^{j2\pi n/N}$ is raised to the Nth power, the result is

$$(e^{j2\pi n/N})^N = e^{j2\pi n} = 1. \qquad 2.33$$

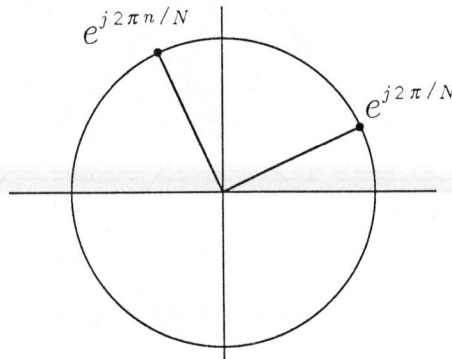

Figure 2.9: The Complex Numbers $e^{j2\pi/N}$ and $e^{j2\pi n/N}$

We say that $e^{j2\pi n/N}$ is one of the Nth roots of unity, meaning that $e^{j2\pi n/N}$ is one of the values of z for which

$$z^N - 1 = 0. \qquad 2.34$$

There are N such roots, namely,

$$e^{j2\pi n/N}, \qquad n = 0, 1, \ldots, N-1. \qquad 2.35$$

As illustrated in Figure 2.10, the 12th roots of unity are uniformly distributed around the unit circle at angles $2\pi n/12$. The sum of all of the Nth roots of unity is zero:

$$S_N = \sum_{n=0}^{N-1} e^{j2\pi n/N} = 0. \qquad 2.36$$

This property, which is obvious from Figure 2.10, is illustrated in Figure 2.11, where the partial sums $S_k = \sum_{n=0}^{k-1} e^{j2\pi n/N}$ are plotted for $k = 1, 2, \ldots, N$.

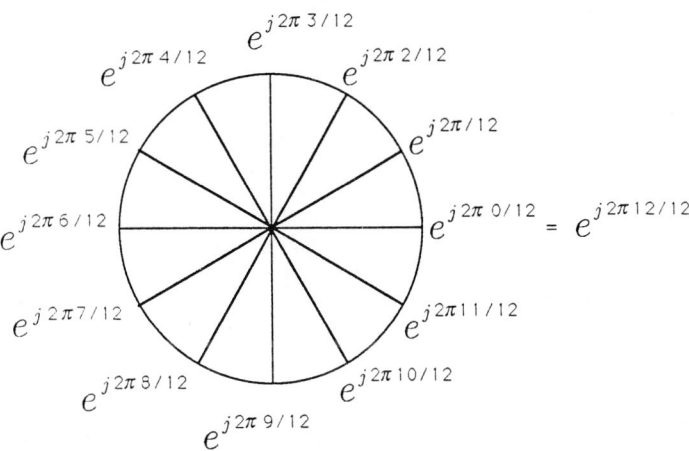

Figure 2.10: Roots of Unity

These partial sums will become important to us in our study of phasors and light diffraction in Chapter 3 and in our discussion of filters in Chapter 6.

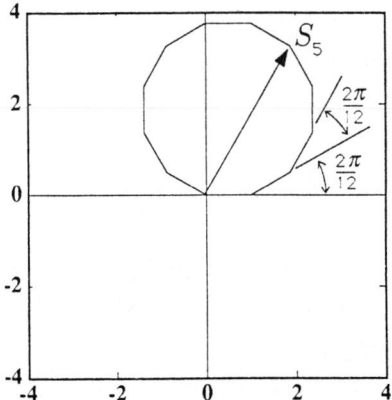

Figure 2.11: Partial Sums of the Roots of Unity

Geometric Sum Formula. It is natural to ask whether there is an analytical expression for the partial sums of roots of unity:

$$S_k = \sum_{n=0}^{k-1} e^{j2\pi n/N}. \tag{2.37}$$

We can imbed this question in the more general question, is there an analytical solution for the "geometric sum"

$$S_k = \sum_{n=0}^{k-1} z^n \quad ?$$

The answer is yes, and here is how we find it. If $z = 1$, the answer is $S_k = k$. If $z \neq 1$, we can premultiply S_k by z and proceed as follows:

$$
\begin{aligned}
zS_k &= \sum_{n=0}^{k-1} z^{n+1} = \sum_{m=1}^{k} z^m \\
&= \sum_{m=0}^{k-1} z^m + z^k - 1 \\
&= S_k + z^k - 1.
\end{aligned}
\tag{2.38}
$$

From this formula we solve for the geometric sum:

$$
S_k = \begin{cases} \frac{1-z^k}{1-z}, & z \neq 1 \\ k, & z = 1. \end{cases}
\tag{2.39}
$$

This basic formula for the geometric sum S_k is used throughout electromagnetic theory and system theory to solve problems in antenna design and spectrum analysis. Never forget it.

Problem 2.13 Find formulas for $S_k = \sum_{n=0}^{k-1} e^{jn\theta}$ and for $S_k = \sum_{n=0}^{k-1} e^{j2\pi/Nn}$. ∎

Problem 2.14 Prove $\sum\limits_{n=0}^{N-1} e^{j2\pi n/N} = 0$. ∎

Problem 2.15 Find formulas for the magnitude and phase of the partial sum $S_k = \sum\limits_{n=0}^{k-1} e^{j2\pi n/N}$. ∎

Problem 2.16 (MATLAB) Write a MATLAB program to compute and plot the partial sum $S_k = \sum\limits_{n=0}^{k-1} e^{j2\pi n/N}$ for $k = 1, 2, \ldots, N$. You should observe Figure 2.11. ∎

Problem 2.17 Solve the equation $(z+1)^3 = z^3$. ∎

Problem 2.18 Find all roots of the equation $z^3 + z^2 + 3z - 15 = 0$. ∎

Problem 2.19 Find c so that $(1+j)$ is a root of the equation $z^{17} + 2z^{15} - c = 0$. ∎

*2.6 Second-Order Differential and Difference Equations

With our understanding of the functions e^x, $e^{j\theta}$, and the quadratic equation $z^2 + \frac{b}{a} z + \frac{c}{a} = 0$, we can undertake a rudimentary study of differential and difference equations.

Differential Equations. In your study of circuits and systems you will encounter the homogeneous differential equation

$$\frac{d^2}{dt^2} x(t) + a_1 \frac{d}{dt} x(t) + a_2 = 0. \qquad 2.40$$

Because the function e^{st} reproduces itself under differentiation, it is plausible to assume that $x(t) = e^{st}$ is a solution to the differential equation. Let's try it:

$$\frac{d^2}{dt^2}(e^{st}) + a_1 \frac{d}{dt}(e^{st}) + a_2(e^{st}) = 0 \qquad 2.41$$

$$(s^2 + a_1 s + a_2)e^{st} = 0.$$

If this equation is to be satisfied for all t, then the polynomial in s must be zero. Therefore we require

$$s^2 + a_1 s + a_2 = 0. \qquad 2.42$$

As we know from our study of this quadratic equation, the solutions are

$$s_{1,2} = -\frac{a_1}{2} \pm \frac{1}{2}\sqrt{a_1^2 - 4a_2}. \qquad 2.43$$

This means that our assumed solution works, provided $s = s_1$ or s_2. It is a fundamental result from the theory of differential equations that the most general solution for $x(t)$ is a linear combination of these assumed solutions:

$$x(t) = A_1 e^{s_1 t} + A_2 e^{s_2 t}. \qquad 2.44$$

If $a_1^2 - 4a_2$ is less than zero, then the roots s_1 and s_2 are complex:

$$s_{1,2} = -\frac{a_1}{2} \pm j\frac{1}{2}\sqrt{4a_2 - a_1^2}. \qquad 2.45$$

Let's rewrite this solution as

$$s_{1,2} = \sigma \pm j\omega \qquad 2.46$$

where σ and ω are the constants

$$\sigma = -\frac{a_1}{2} \qquad 2.47$$

$$\omega = \frac{1}{2}\sqrt{4a_2 - a_1^2}. \qquad 2.48$$

With this notation, the solution for $x(t)$ is

$$x(t) = A_1 e^{\sigma t} e^{j\omega t} + A_2 e^{\sigma t} e^{-j\omega t}. \qquad 2.49$$

If this solution is to be real, then the two terms on the right-hand side must be complex conjugates. This means that $A_2 = A_1^*$ and the solution for $x(t)$ is

$$\begin{aligned} x(t) &= A_1 e^{\sigma t} e^{j\omega t} + A_1^* e^{\sigma t} e^{-j\omega t} \\ &= 2\text{Re}\{A_1 e^{\sigma t} e^{j\omega t}\}. \end{aligned} \qquad 2.50$$

The constant A_1 may be written as $A_1 = |A| e^{j\phi}$. Then the solution for $x(t)$ is

$$x(t) = 2|A| e^{\sigma t} \cos(\omega t + \phi). \qquad 2.51$$

This "damped cosinusoidal solution" is illustrated in Figure 2.12.

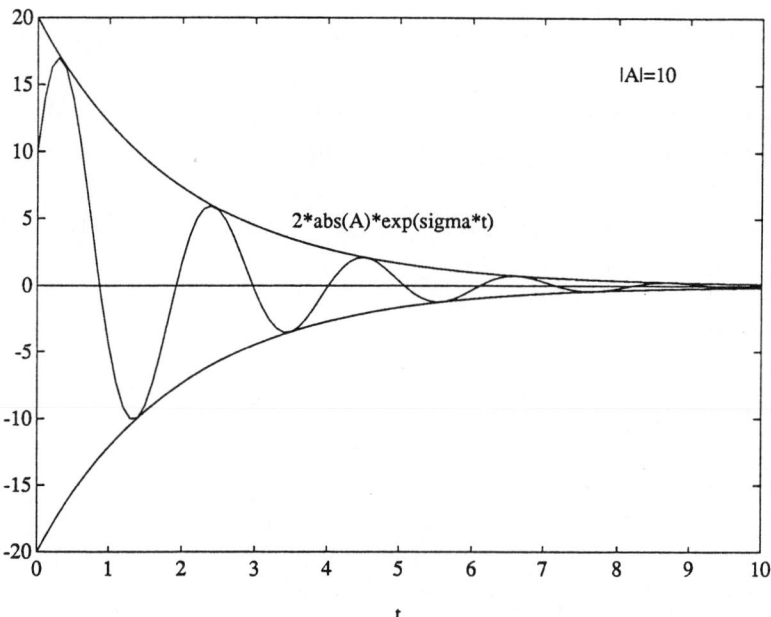

Figure 2.12: The Solution to a Second-Order Differential Equation

Problem 2.20   Find the general solutions to the following differential equations:

(a) $\frac{d^2}{dt^2}x(t) + 2\frac{d}{dt}x(t) + 2 = 0$;

(b) $\frac{d^2}{dt^2}x(t) - 2\frac{d}{dt}x(t) + 2 = 0$;

(c) $\frac{d^2}{dt^2}x(t) + 2 = 0$. ∎

Difference Equations. In your study of digital filters you will encounter homogeneous difference equations of the form

$$x_n + a_1 x_{n-1} + a_2 x_{n-2} = 0. \qquad 2.52$$

What this means is that the *sequence* $\{x_n\}$ obeys a *homogeneous recursion*:

$$x_n = -a_1 x_{n-1} - a_2 x_{n-2}.$$

A plausible guess at a solution is the geometric sequence $x_n = z^n$. With this guess, the difference equation produces the result

$$z^n + a_1 z^{n-1} + a_2 z^{n-2} = 0 \qquad 2.53$$

$$(1 + a_1 z^{-1} + a_2 z^{-2})z^n = 0.$$

If this guess is to work, then the second-order polynomial on the left-hand side must equal zero:

$$1 + a_1 z^{-1} + a_2 z^{-2} = 0 \qquad 2.54$$

$$z^2 + a_1 z + a_2 = 0.$$

The solutions are

$$z_{1,2} = -\frac{a_1}{2} \pm j\frac{1}{2}\sqrt{4a_2 - a_1^2} \qquad 2.55$$

$$= re^{j\theta}.$$

The general solution to the difference equation is a linear combination of the assumed solutions:

$$
\begin{aligned}
x_n &= A_1 z_1^n + A_2 (z_1^*)^n \\
&= A_1 z_1^n + A_1^* (z_1^*)^n \\
&= 2\mathrm{Re}\{A_1 z_1^n\} \\
&= 2|A| r^n \cos(\theta n + \phi).
\end{aligned}
\qquad 2.56
$$

This general solution is illustrated in Figure 2.13.

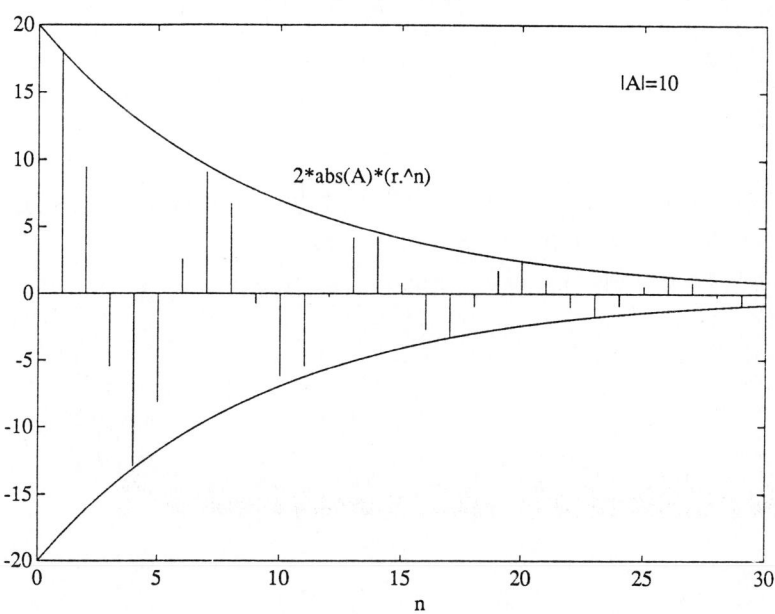

Figure 2.13: The Solution to a Second-Order Difference Equation

Problem 2.21 Find the general solutions to the following difference equations:

  (a) $x_n + 2x_{n-1} + 2 = 0$;

  (b) $x_n - 2x_{n-1} + 2 = 0$;

  (c) $x_n + 2x_{n-2} = 0$. ∎

2.7 Numerical Experiment (Approximating $e^{j\theta}$)

We have demonstrated that the function $e^{j\theta}$ has two representations:

(1) $e^{j\theta} = \lim\limits_{n\to\infty} \left(1 + \frac{j\theta}{n}\right)^n$; and

(2) $e^{j\theta} = \lim\limits_{n\to\infty} \sum\limits_{k=0}^{n} \frac{(j\theta)^k}{k!}$.

In this experiment, you will write a MATLAB program to evaluate the two functions f_n and S_n for twenty values of n:

(1) $f_n = \left(1 + \frac{j\theta}{n}\right)^n$, $n = 1, 2, \ldots, 20$; and

(2) $S_n = \sum\limits_{k=0}^{n} \frac{(j\theta)^k}{k!}$, $n = 1, 2, \ldots, 20$.

Choose $\theta = \pi/4$ ($=$ `pi/4`). Use an implicit `for` loop to draw and plot a circle of radius 1. Then use an implicit `for` loop to compute and plot f_n and an explicit `for` loop to compute and plot S_n for $n = 1, 2, \ldots, 100$. You should observe plots like those illustrated in Figure 2.14. Interpret them.

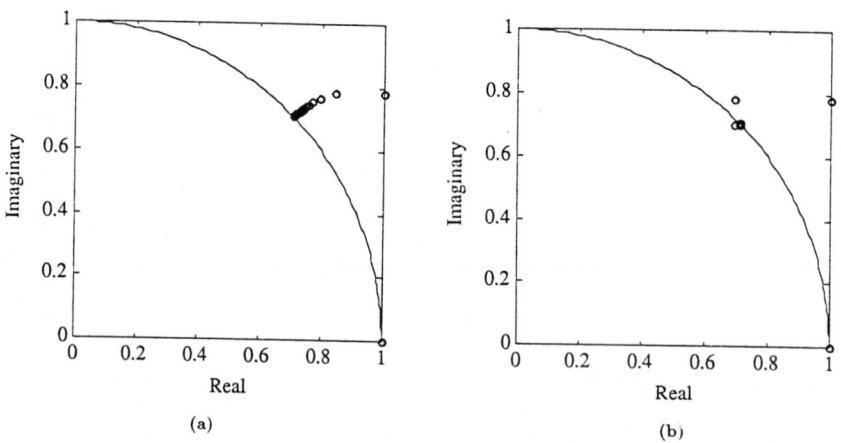

Figure 2.14: Plots for Convergence; (a) of f_n, and (b) of S_n

3

| | | | | | | | | | | | | **3** | |

Phasors

Notes to Teachers and Students:

Phasors! For those who understand them, they are of incomparable value for the study of elementary and advanced topics. For those who misunderstand them, they are a constant source of confusion and are of no apparent use. So, let's understand them.

The conceptual leap from the complex number $e^{j\theta}$ to the phasor $e^{j(\omega t+\theta)}$ comes in Section 3.2. Take as long as necessary to understand every geometrical and algebraic nuance. Write the MATLAB program in Problem 3.6 to fix the key ideas once and for all. Then *use* phasors to study beating between tones, multiphase power, and Lissajous figures in Sections 3.3 through 3.5. We usually conduct a classroom demonstration of beating between tones using two phase-locked sources, an oscilloscope, and a speaker. We also demonstrate Lissajous figures with this hardware.

Sections 3.6 and 3.7 on sinusoidal steady state and light scattering are too demanding for freshmen but are right on target for sophomores. These sections may be covered in a sophomore course (or a supplement to a sophomore course) or skipped in a freshman course without consequence.

In the numerical experiment in Section 3.8, students compute and plot interference patterns for two sinusoids that are out of phase.

3.1 Introduction

In engineering and applied science, three test signals form the basis for our study of electrical and mechanical systems. The *impulse* is an idealized signal that models very short excitations (like current pulses, hammer blows, pile drives, and light flashes). The *step* is an idealized signal that models excitations that are switched on and stay on (like current in a relay that closes or a transistor that switches). The *sinusoid* is an idealized signal that models excitations that oscillate with a regular frequency (like AC power, AM radio, pure musical tones, and harmonic vibrations). All three signals are used in the laboratory to design and analyze electrical and mechanical circuits, control systems, radio antennas, and the like. The sinusoidal signal is particularly important because it may be used to determine the frequency selectivity of a circuit (like a superheterodyne radio receiver) to excitations of different frequencies. For this reason, every manufacturer of electronics test equipment builds sinusoidal oscillators that may be swept through many octaves of frequency. (Hewlett-Packard was started in 1940 with the famous HP audio oscillator.)

In this chapter we use what we have learned about complex numbers and the function $e^{j\theta}$ to develop a *phasor calculus* for representing and manipulating sinusoids. This calculus operates very much like the calculus we developed in Chapters 1 and 2 for manipulating complex numbers. We apply our calculus to the study of beating phenomena, multiphase power, series RLC circuits, and light scattering by a slit.

3.2 Phasor Representation of Signals

There are two key ideas behind the phasor representation of a signal:

(i) a real, time-varying signal may be represented by a complex, time-varying signal; and

(ii) a complex, time-varying signal may be represented as the product of a complex number that is *independent* of time and a complex signal that is *dependent* on time.

Let's be concrete. The signal

$$x(t) = A\cos(\omega t + \phi), \qquad\qquad 3.1$$

illustrated in Figure 3.1, is a cosinusoidal signal with amplitude A, frequency ω, and phase ϕ. The amplitude A characterizes the peak-to-peak swing of $2A$, the angular frequency ω characterizes the period $T = \frac{2\pi}{\omega}$ between negative-to-positive zero crossings (or positive peaks or negative peaks), and the phase ϕ characterizes the time $\tau = \frac{-\phi}{\omega}$ when the signal reaches its first peak. With τ so defined, the signal $x(t)$ may also be written as

$$x(t) = A\cos\omega(t - \tau). \qquad\qquad 3.2$$

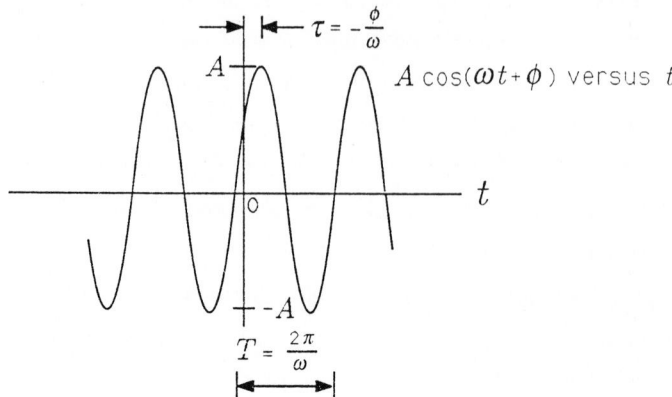

Figure 3.1: A Cosinusoidal Signal

When τ is positive, then τ is a "time delay" that describes the time (greater than zero) when the first peak is achieved. When τ is negative, then τ is a "time advance" that describes the time (less than zero) when the last peak was achieved. With the substitution $\omega = \frac{2\pi}{T}$, we obtain a third way of writing $x(t)$:

$$x(t) = A\cos\frac{2\pi}{T}(t - \tau).$$

3.3

In this form the signal is easy to plot. Simply draw a cosinusoidal wave with amplitude A and period T; then strike the origin ($t = 0$) so that the signal reaches its peak at τ. In summary, the parameters that determine a cosinusoidal signal have the following units:

A, arbitrary (e.g., volts or meters/sec, depending upon the application)

ω, in radians/sec (rad/sec)

T, in seconds (sec)

ϕ, in radians (rad)

τ, in seconds (sec)

Problem 3.1 Show that $x(t) = A\cos\frac{2\pi}{T}(t - \tau)$ is "periodic with period T," meaning that $x(t + mT) = x(t)$ for all integer m. ∎

Problem 3.2 The inverse of the period T is called the "temporal frequency" of the cosinusoidal signal and is given the symbol f; the units of $f = \frac{1}{T}$ are (seconds)$^{-1}$ or hertz (Hz). Write $x(t)$ in terms of f. How is f related to ω? Explain why f gives the number of cycles of $x(t)$ per second. ∎

Problem 3.3 Sketch the function $x(t) = 110\cos\left[2\pi(60)t - \frac{\pi}{8}\right]$ versus t. Repeat for $x(t) = 5\cos\left[2\pi(16 \times 10^6)t + \frac{\pi}{4}\right]$ and $x(t) = 2\cos\left[\frac{2\pi}{10^{-3}}\left(t - \frac{10^{-3}}{8}\right)\right]$. For each function, determine A, ω, T, f, ϕ, and τ. Label your sketches carefully. ∎

The signal $x(t) = A\cos(\omega t + \phi)$ can be *represented* as the real part of a complex number:

$$x(t) = \text{Re}[Ae^{j(\omega t + \phi)}]$$
$$= \text{Re}[Ae^{j\phi}e^{j\omega t}].$$

3.4

We call $Ae^{j\phi}e^{j\omega t}$ the complex representation of $x(t)$ and write

$$x(t) \quad \leftrightarrow \quad Ae^{j\phi}e^{j\omega t},$$

3.5

meaning that the signal $x(t)$ may be reconstructed by taking the real part of $Ae^{j\phi}e^{j\omega t}$. In this representation, we call $Ae^{j\phi}$ the *phasor* or complex amplitude representation of $x(t)$ and write

$$x(t) \quad \leftrightarrow \quad Ae^{j\phi},$$

3.6

meaning that the signal $x(t)$ may be reconstructed from $Ae^{j\phi}$ by multiplying with $e^{j\omega t}$ and taking the real part. In communication theory, we call $Ae^{j\phi}$ the *baseband* representation of the signal $x(t)$.

Problem 3.4 For each of the signals in Problem 3.3, give the corresponding phasor representation $Ae^{j\phi}$. ∎

Geometric Interpretation. Let's call

$$Ae^{j\phi}e^{j\omega t}$$

3.7

the complex representation of the real signal $A\cos(\omega t + \phi)$. At $t = 0$, the complex representation produces the *phasor*

$$Ae^{j\phi}. \tag{3.8}$$

This phasor is illustrated in Figure 3.2. In the figure, ϕ is approximately $\frac{-\pi}{10}$. If we let t increase to time t_1, then the complex representation produces the phasor

$$Ae^{j\phi}e^{j\omega t_1}. \tag{3.9}$$

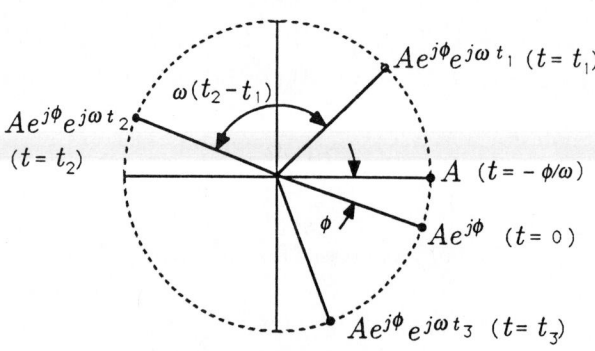

Figure 3.2: Rotating Phasor

We know from our study of complex numbers that $e^{j\omega t_1}$ just rotates the phasor $Ae^{j\phi}$ through an angle of ωt_1! See Figure 3.2. Therefore, as we run t from 0, indefinitely, we rotate the phasor $Ae^{j\phi}$ indefinitely, turning out the circular trajectory of Figure 3.2. When $t = \frac{2\pi}{\omega}$, then $e^{j\omega t} = e^{j2\pi} = 1$. Therefore, every $\left(\frac{2\pi}{\omega}\right)$ seconds, the phasor revisits any given position on the circle of radius A. We sometimes call $Ae^{j\phi}e^{j\omega t}$ a *rotating phasor* whose rotation rate is the frequency ω:

$$\frac{d}{dt}\omega t = \omega. \tag{3.10}$$

This rotation rate is also the frequency of the cosinusoidal signal $A\cos(\omega t + \phi)$.

In summary, $Ae^{j\phi}e^{j\omega t}$ is the complex, or rotating phasor, representation of the signal $A\cos(\omega t + \phi)$. In this representation, $e^{j\omega t}$ rotates the phasor $Ae^{j\phi}$ through angles ωt at the rate ω. The real part of the complex representation is the desired signal $A\cos(\omega t + \phi)$. This real part is read off the rotating phasor diagram as illustrated in Figure 3.3. In the figure, the angle ϕ is about $-\frac{2\pi}{10}$. As we become more facile with phasor representations, we will write $x(t) = \text{Re}[Xe^{j\omega t}]$ and call $Xe^{j\omega t}$ the complex representation and X the phasor representation. The phasor X is, of course, just the phasor $Ae^{j\phi}$.

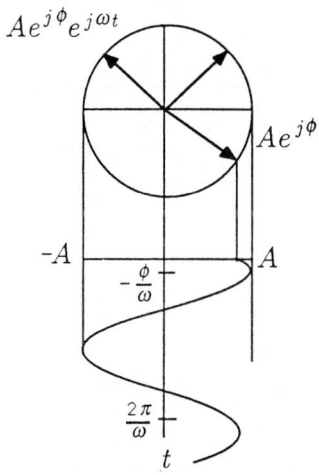

Figure 3.3: Reading a Real Signal from a Complex, Rotating Phasor

Problem 3.5   Sketch the imaginary part of $Ae^{j\phi}e^{j\omega t}$ to show that this is $A\sin(\omega t + \phi)$. What do we mean when we say that the real and imaginary parts of $Ae^{j\phi}e^{j\omega t}$ are "90° out of phase"? ∎

Problem 3.6 (MATLAB) Modify Demo 2.1 in Chapter 2 so that $\theta = \omega t$, with ω an input frequency variable and t a time variable that ranges from $-2\left(\frac{2\pi}{\omega}\right)$ to $+2\left(\frac{2\pi}{\omega}\right)$ in steps of $0.02\left(\frac{2\pi}{\omega}\right)$. In your modified program, compute and plot $e^{j\omega t}$, $\text{Re}[e^{j\omega t}]$, and $\text{Im}[e^{j\omega t}]$ for $-2\left(\frac{2\pi}{\omega}\right) \le t < 2\left(\frac{2\pi}{\omega}\right)$ in steps of $0.02\left(\frac{2\pi}{\omega}\right)$. Plot $e^{j\omega t}$ in a two-dimensional plot to get a picture like Figure 3.2 and plot $\text{Re}[e^{j\omega t}]$ and $\text{Im}[e^{j\omega t}]$ versus t to get signals like those of Figure 3.1. You should observe something like Figure 3.4 using the subplot features discussed in Appendix A. (In the figure, w represents Greek ω.) ∎

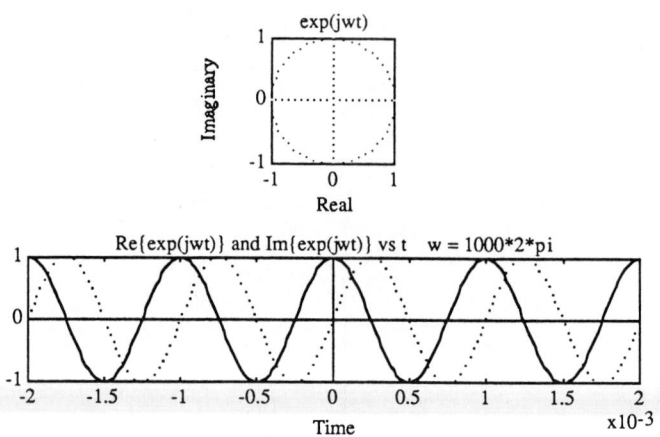

Figure 3.4: The Functions $e^{j\omega t}$, Re$[e^{j\omega t}]$, and Im$[e^{j\omega t}]$

Positive and Negative Frequencies. There is an alternative phasor representation for the signal $x(t) = A\cos(\omega t + \phi)$. We obtain it by using the Euler formula of Chapter 2, namely, $\cos\theta = \frac{1}{2}(e^{j\theta} + e^{-j\theta})$. When this formula is applied to $x(t)$, we obtain the result

$$x(t) = \frac{A}{2}[e^{j(\omega t + \phi)} + e^{-j(\omega t + \phi)}]$$
$$= \frac{A}{2}e^{j\phi}e^{j\omega t} + \frac{A}{2}e^{-j\phi}e^{-j\omega t}.$$

3.11

In this formula, the term $\frac{A}{2}e^{j\phi}e^{j\omega t}$ is a rotating phasor that begins at the phasor value $Ae^{j\phi}$ (for $t = 0$) and rotates counterclockwise with frequency ω. The term $\frac{A}{2}e^{-j\phi}e^{-j\omega t}$ is a rotating phasor that begins at the (complex conjugate) phasor value $\frac{A}{2}e^{-j\phi}$ (for $t = 0$) and rotates clockwise with (negative) frequency ω. The physically meaningful frequency for a cosine is ω, a positive number like $2\pi(60)$ for 60 Hz power. There is no such thing as a negative frequency. The so-called negative frequency of the term $\frac{A}{2}e^{-j\phi}e^{-j\omega t}$ just indicates that the direction of rotation for the rotating phasor is clockwise and not counterclockwise. The notion of a negative frequency is just an artifact of the two-phasor representation of $A\cos(\omega t + \phi)$. In the one-phasor representation, when we take the "real part," the artifact does not arise. In your study of circuits, systems theory, electromagnetics, solid-state devices, signal processing, control, and communications, you will encounter both the one- and two-phasor representations. Become facile with them.

Problem 3.7   Sketch the two-phasor representation of $A\cos(\omega t + \phi)$. Show clearly how this representation works by discussing the counterclockwise rotation of the positive frequency part and the clockwise rotation of the negative frequency part. ∎

Adding Phasors. The sum of two signals with common frequencies but different amplitudes and phases is

$$A_1\cos(\omega t + \phi_1) + A_2\cos(\omega t + \phi_2).\qquad 3.12$$

The rotating phasor representation for this sum is

$$(A_1 e^{j\phi_1} + A_2 e^{j\phi_2})e^{j\omega t}.\qquad 3.13$$

The new phasor is $A_1 e^{j\phi_1} + A_2 e^{j\phi_2}$, and the corresponding real signal is $x(t) = \mathrm{Re}\big[(A_1 e^{j\phi_1} + A_2 e^{j\phi_2})e^{j\omega t}\big]$. The new phasor is illustrated in Figure 3.5.

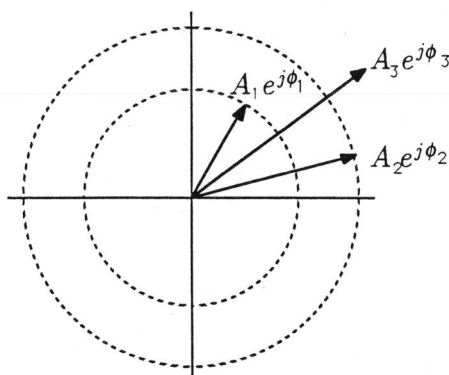

Figure 3.5: Adding Phasors

Problem 3.8 Write the phasor $A_1 e^{j\phi_1} + A_2 e^{j\phi_2}$ as $A_3 e^{j\phi_3}$; determine A_3 and ϕ_3 in terms of A_1, A_2, ϕ_1, and ϕ_2. What is the corresponding real signal? ∎

Differentiating and Integrating Phasors. The derivative of the signal $A \cos(\omega t + \phi)$ is the signal

$$\begin{aligned}
\frac{d}{dt} A \cos(\omega t + \phi) &= -\omega A \, \sin(\omega t + \phi) \\
&= -\mathrm{Im}[\omega A e^{j\phi} e^{j\omega t}] \\
&= \mathrm{Re}[j\omega A e^{j\phi} e^{j\omega t}] \\
&= \mathrm{Re}[\omega e^{j\pi/2} A e^{j\phi} e^{j\omega t}].
\end{aligned} \qquad 3.14$$

This finding is very important. It says that the derivative of $A \cos(\omega t + \phi)$ has the phasor representation

$$\begin{aligned}
\frac{d}{dt} A \cos(\omega t + \phi) \quad &\leftrightarrow \quad j\omega A e^{j\phi} \\
&\leftrightarrow \quad \omega e^{j\pi/2} A e^{j\phi}.
\end{aligned} \qquad 3.15$$

These two phasor representations are entirely equivalent. The first says that the phasor $A e^{j\phi}$ is complex scaled by $j\omega$ to produce the phasor for $\frac{d}{dt} A \cos(\omega t + \phi)$, and the second says that it is scaled by ω and phased by $+\pi/2$. The phasor representations of $A \cos(\omega t + \phi)$ and $\frac{d}{dt} A \cos(\omega t + \phi)$ are illustrated in Figure 3.6. Note that the derivative "leads by $\pi/2$ radians (90°)."

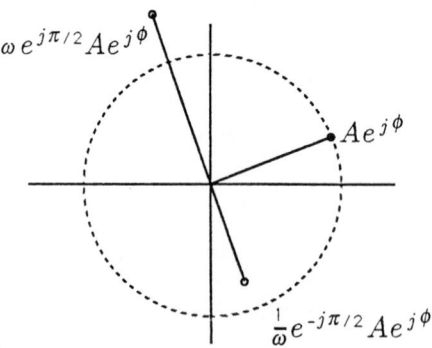

Figure 3.6: Differentiating and Integrating Phasors

The integral of $A\cos(\omega t + \phi)$ is

$$\int A\cos(\omega t + \phi)\,dt = \frac{A}{\omega}\sin(\omega t + \phi)$$

$$= \mathrm{Im}\left[\frac{A}{\omega}e^{j\phi}e^{j\omega t}\right]$$

$$= \mathrm{Re}\left[-j\frac{A}{\omega}e^{j\phi}e^{j\omega t}\right] \qquad 3.16$$

$$= \mathrm{Re}\left[\frac{A}{j\omega}e^{j\phi}e^{j\omega t}\right]$$

$$= \mathrm{Re}\left[\frac{1}{\omega}e^{-j\pi/2}Ae^{j\phi}e^{j\omega t}\right].$$

This finding shows that the integral of $A\cos(\omega t + \phi)$ has the phasor representation

$$\int A\cos(\omega t + \phi)\,dt \quad \leftrightarrow \quad \frac{1}{j\omega}Ae^{j\phi}$$

$$\qquad 3.17$$

$$\leftrightarrow \quad \frac{1}{\omega}e^{-j\pi/2}Ae^{j\phi}.$$

The phasor $Ae^{j\phi}$ is complex scaled by $\frac{1}{j\omega}$ or scaled by $\frac{1}{\omega}$ and phased by $e^{-j\pi/2}$ to produce the phasor for $\int A\cos(\omega t + \phi)\,dt$. This is illustrated in Figure 3.6. Note that the integral "lags by $\pi/2$ radians ($90°$)." Keep these geometrical pictures of leading and lagging by $\pi/2$ in your mind at all times as you continue your more advanced study of engineering.

An Aside: The Harmonic Oscillator. The signal $A\cos(\omega t + \phi)$ stands on its own as an interesting signal. But the fact that it reproduces itself (with scaling and phasing) under differentiation means that it obeys the second-order differential equation of the simple harmonic oscillator.[1] That is, the differential equation

$$\frac{d^2x(t)}{dt^2} + \omega^2 x(t) = 0 \qquad 3.18$$

has the solution

$$x(t) = A\cos(\omega t + \phi). \qquad 3.19$$

---

[1] This means, also, that we have an easy way to synthesize cosines with circuits that obey the equation of a simple harmonic oscillator!

Try it:

$$\frac{d^2}{dt^2} x(t) = \frac{d}{dt}\left[-A\omega \sin(\omega t + \phi)\right] = -\omega^2 A \cos(\omega t + \phi). \qquad 3.20$$

The constants A and ϕ are determined from the initial conditions

$$x(0) = A \cos \phi \qquad x^2(0) + x^2\left(\frac{\pi}{2\omega}\right) = A^2$$

$$\Longleftrightarrow \qquad\qquad\qquad 3.21$$

$$x\left(\frac{\pi}{2\omega}\right) = -A\sin\phi \qquad -\frac{x(\pi/2\omega)}{x(0)} = \tan\phi.$$

Problem 3.9 Show how to compute A and ϕ in the equation $x(t) = A\cos(\omega t + \phi)$ from the initial conditions $x(0)$ and $\frac{d}{dt} x(t)\big|_{t=0}$. ∎

3.3 Beating between Tones

Perhaps you have heard two slightly mistuned musical instruments play pure tones whose frequencies are close but not equal. If so, you have sensed a beating phenomenon wherein a pure tone seems to wax and wane. This waxing and waning tone is, in fact, a tone whose frequency is the average of the two mismatched frequencies, amplitude modulated by a tone whose "beat" frequency is half the difference between the two mismatched frequencies. The effect is illustrated in Figure 3.7. Let's see if we can derive a mathematical model for the beating of tones.

We begin with two pure tones whose frequencies are $\omega_0 + \nu$ and $\omega_0 - \nu$ (for example, $\omega_0 = 2\pi \times 10^3$ rad/sec and $\nu = 2\pi$ rad/sec). The average frequency is ω_0, and the difference frequency is 2ν. What you hear is the sum of the two tones:

$$x(t) = A_1 \cos\left[(\omega_0 + \nu)t + \phi_1\right] + A_2 \cos\left[(\omega_0 - \nu)t + \phi_2\right]. \qquad 3.22$$

The first tone has amplitude A_1 and phase ϕ_1; the second has amplitude A_2 and phase ϕ_2. We will assume that the two amplitudes are equal to A. Furthermore, whatever the phases, we may write them as

$$\phi_1 = \phi + \psi \text{ and } \phi_2 = \phi - \psi \qquad\qquad 3.23$$

$$\phi = \frac{1}{2}(\phi_1 + \phi_2) \text{ and } \psi = \frac{1}{2}(\phi_1 - \phi_2).$$

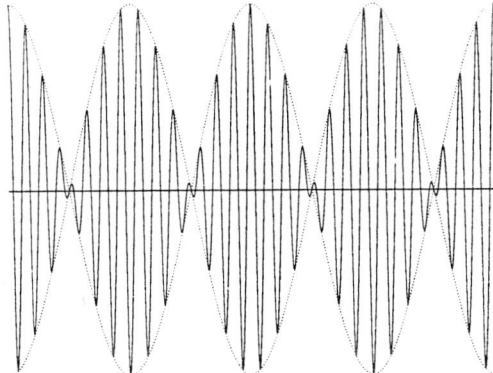

Figure 3.7: Beating between Tones

Recall our trick for representing $x(t)$ as a complex phasor:

$$
\begin{aligned}
x(t) &= A\operatorname{Re}\{e^{j[(\omega_0+\nu)t+\phi+\psi]} + e^{j[(\omega_0-\nu)t+\phi-\psi]}\}\\
&= A\operatorname{Re}\{e^{j(\omega_0 t+\phi)}[e^{j(\nu t+\psi)} + e^{-j(\nu t+\psi)}]\}\\
&= 2A\operatorname{Re}\{e^{j(\omega_0 t+\phi)}\cos(\nu t+\psi)\}\\
&= 2A\,\cos(\omega_0 t+\phi)\cos(\nu t+\psi).
\end{aligned}
\qquad 3.24
$$

This is an amplitude modulated wave, wherein a low frequency signal with beat frequency ν rad/sec modulates a high frequency signal with carrier frequency ω_0 rad/sec. Over short periods of time, the modulating term $\cos(\nu t + \psi)$ remains essentially constant while the carrier term $\cos(\omega_0 t + \phi)$ turns out many cycles of its tone. For example, if t runs from 0 to $\frac{2\pi}{10\nu}$ (about 0.1 seconds in our example), then the modulating wave turns out just 1/10 cycle while the carrier turns out $\frac{\omega_0}{10\nu}$ cycles (about 100 in our example). Every time νt changes by 2π radians, then the modulating term goes from a maximum (a wax) through a minimum (a wane) and back to a maximum. This cycle takes

$$\nu t = 2\pi \quad\Longleftrightarrow\quad t = \frac{2\pi}{\nu} \text{ seconds,} \qquad\qquad 3.25$$

which is 1 second in our example. In this 1 second the carrier turns out 1000 cycles.

Problem 3.10   Find out the frequency of A above middle C on a piano. Assume two pianos are mistuned by ± 1 Hz ($\pm 2\pi$ rad/sec). Find their beat frequency ν and their carrier frequency ω_0. ∎

Problem 3.11   (MATLAB) Write a MATLAB program to compute and plot $A \cos[(\omega_0 + \nu)t + \phi_1]$, $A \cos[(\omega_0 - \nu)t + \phi_2]$, and their sum. Then compute and plot $2A \cos(\omega_0 t + \phi) \cos(\nu t + \psi)$. Verify that the sum equals this latter signal. ∎

3.4 Multiphase Power

The electrical service to your home is a two-phase service.[2] This means that two 110 volt, 60 Hz lines, plus neutral, terminate in the panel. The lines are π radians (180°) out of phase, so we can write them as

$$x_1(t) = 110 \cos\left[2\pi(60)t + \phi\right] = \text{Re}\left\{110e^{j[2\pi(60)t+\phi]}\right\}$$
$$= \text{Re}\left\{X_1 e^{j2\pi(60)t}\right\}$$

$$X_1 = 110e^{j\phi} \qquad\qquad 3.26$$

$$x_2(t) = 110 \cos\left[2\pi(60)t + \phi + \pi\right] = \text{Re}\left\{110e^{j[2\pi(60)t+\phi+\pi]}\right\}$$
$$= \text{Re}\left\{X_2 e^{j2\pi(60)t}\right\}$$

$$X_2 = 110e^{j(\phi+\pi)}.$$

These two voltages are illustrated as the phasors X_1 and X_2 in Figure 3.8.

---

[2] It really is, although it is said to be "single phase" because of the way it is picked off a single phase of a primary source. You will hear more about this in circuits and power courses.

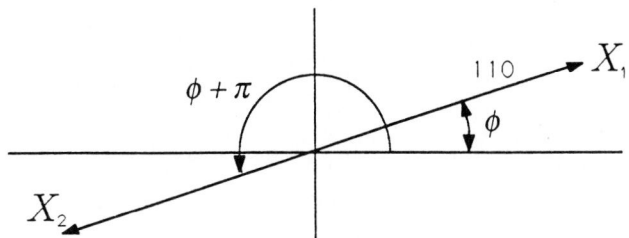

Figure 3.8: Phasors in Two-Phase Power

You may use $x_1(t)$ to drive your clock radio or your toaster and the difference between $x_1(t)$ and $x_2(t)$ to drive your range or dryer:

$$x_1(t) - x_2(t) = 220 \cos\left[2\pi(60)t + \phi\right]. \qquad 3.27$$

The phasor representation of this difference is

$$X_1 - X_2 = 220e^{j\phi}. \qquad 3.28$$

The breakers in a breaker box span the x_1-to-neutral bus for 110 volts and the x_1-to-x_2 buses for 220 volts.

Problem 3.12 Sketch the phasor $X_1 - X_2$ on Figure 3.8. ∎

Most industrial installations use a three-phase service consisting of the signals $x_1(t)$, $x_2(t)$, and $x_3(t)$:

$$x_n(t) = 110\mathrm{Re}\{e^{j[\omega_0 t + n(2\pi/3)]}\} \quad \leftrightarrow \quad X_n = 110e^{jn(2\pi/3)}, \quad n = 1, 2, 3. \quad 3.29$$

The phasors for three-phase power are illustrated in Figure 3.9.

Problem 3.13 Sketch the phasor $X_2 - X_1$ corresponding to $x_2(t) - x_1(t)$ on Figure 3.9. Compute the voltage you can get with $x_2(t) - x_1(t)$. This answer explains why you do not get 220 volts in three-phase circuits. What do you get? ∎

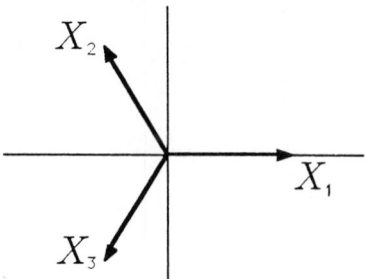

Figure 3.9: Three-Phase Power

Constant Power. Two- and three-phase power generalizes in an obvious way to N-phase power. In such a scheme, the N signals x_n ($n = 0, 1, \ldots, N-1$) are

$$x_n(t) = A \cos\left(\omega t + \frac{2\pi}{N} n\right)$$
$$= \mathrm{Re}[A e^{j2\pi n/N} e^{j\omega t}] \quad \leftrightarrow \quad X_n = A e^{j2\pi n/N}. \tag{3.30}$$

The phasors X_n are $A e^{j2\pi(n/N)}$. The sum of all N signals is zero:

$$\sum_{n=0}^{N-1} x_n(t) = \mathrm{Re}\left\{ A \sum_{n=0}^{N-1} e^{j2\pi n/N} e^{j\omega t} \right\}$$
$$= \mathrm{Re}\left\{ A \frac{1 - e^{j2\pi}}{1 - e^{j2\pi/N}} e^{j\omega t} \right\} \tag{3.31}$$
$$= 0.$$

But what about the sum of the instantaneous powers? Define the instantaneous power of the nth signal to be

$$p_n(t) = x_n^2(t) = A^2 \cos^2\left(\omega t + \frac{2\pi}{N} n\right)$$
$$= \frac{A^2}{2} + \frac{A^2}{2} \cos\left(2\omega t + 2 \frac{2\pi}{N} n\right) \tag{3.32}$$
$$= \frac{A^2}{2} + \mathrm{Re}\left\{ \frac{A^2}{2} e^{j(2\pi/N)2n} e^{j2\omega t} \right\}.$$

The sum of all instantaneous powers is (see Problem 3.14)

$$P = \sum_{n=0}^{N-1} p_n(t) = N\frac{A^2}{2},$$ 3.33

and this is independent of time!

Problem 3.14   Carry out the computations of Equation 3.33 to prove that instantaneous power P is constant in the N-phase power scheme. ∎

3.5 Lissajous Figures

Lissajous figures are figures that are turned out on the face of an oscilloscope when sinusoidal signals with different amplitudes and different phases are applied to the time base (real axis) and deflection plate (imaginary axis) of the scope. The electron beam that strikes the phosphorous face then has position

$$z(t) = A_x \cos(\omega t + \phi_x) + jA_y \cos(\omega t + \phi_y).$$ 3.34

In this representation, $A_x \cos(\omega t + \phi_x)$ is the "x-coordinate of the point," and $A_y \cos(\omega t + \phi)$ is the "y-coordinate of the point." As time runs from 0 to infinity, the point $z(t)$ turns out a trajectory like that of Figure 3.10. The figure keeps overwriting itself because $z(t)$ repeats itself every $\frac{2\pi}{\omega}$ seconds. Do you see why?

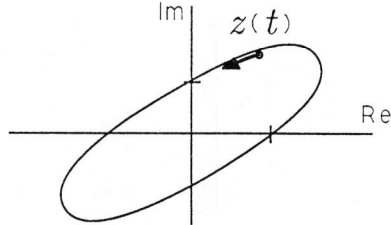

Figure 3.10: Lissajous Figure on Oscilloscope Screen

Problem 3.15 Find the intercepts that the Lissajous figure makes with the real and imaginary axes in Figure 3.10. At what values of time are these intercepts made? ∎

Problem 3.16 Show that the Lissajous figure $z(t) = A_x \cos(\omega t + \phi_x) + j A_y \cos(\omega t + \phi_y)$ is just the rotating phasor $A e^{j(\omega t + \phi)}$ when $A_x = A_y = A$, $\phi_x = \phi$, and $\phi_y = \phi + \frac{\pi}{2}$. ∎

Two-Phasor Representation. We gain insight into the shape of the Lissajous figure if we use Euler's formulas to write $z(t)$ as follows:

$$
z(t) = \frac{A_x}{2}\left[e^{j(\omega t + \phi_x)} + e^{-j(\omega t + \phi_x)}\right] + j\frac{A_y}{2}\left[e^{j(\omega t + \phi_y)} + e^{-j(\omega t + \phi_y)}\right]
$$
$$
= \left[\frac{A_x e^{j\phi_x} + j A_y e^{j\phi_y}}{2}\right]e^{j\omega t} + \left[\frac{A_x e^{-j\phi_x} + j A_y e^{-j\phi_y}}{2}\right]e^{-j\omega t}.
$$

3.35

This representation is illustrated in Figure 3.11. It consists of two *rotating* phasors, with respective phasors B_1 and B_2:

$$
z(t) = B_1 e^{j\omega t} + B_2 e^{-j\omega t}
$$

3.36

$$
B_1 = \frac{A_x e^{j\phi_x} + j A_y e^{j\phi_y}}{2}
$$

$$
B_2 = \frac{A_x e^{-j\phi_x} + j A_y e^{-j\phi_y}}{2}.
$$

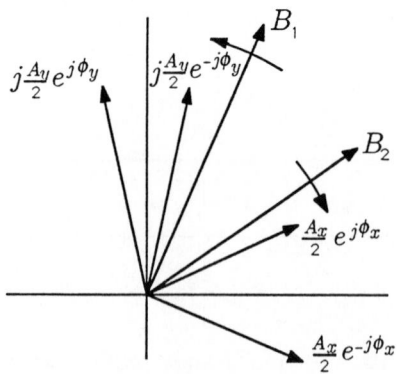

Figure 3.11: Two-Phasor Representation of a Lissajous Figure

As t increases, the phasors rotate past each other where they constructively add to produce large excursions of $z(t)$ from the origin, and then they rotate to antipodal positions where they destructively add to produce near approaches of $z(t)$ to the origin.

In electromagnetics and optics, the representations of $z(t)$ given in Equations 3.34 and 3.36 are called, respectively, *linear* and *circular* representations of elliptical polarization. In the linear representation, the x- and y-components of z vary along the horizontal and vertical lines. In the circular representation, two phasors rotate in opposite directions to turn out circular trajectories whose sum produces the same effect.

Problem 3.17 (MATLAB) Write a MATLAB program to compute and plot the Lissajous figure $z(t)$ when $A_x = 1/2$, $A_y = 1$, $\phi_x = 0$, and $\phi_y = \pi/6$. Discretize t appropriately and choose an appropriate range of values for t. ∎

*3.6 Sinusoidal Steady State and the Series *RLC* Circuit

Phasors may be used to analyze the behavior of electrical and mechanical systems that have reached a kind of equilibrium called *sinusoidal steady state*. In the sinusoidal steady state, every voltage and current (or force and velocity) in a system is sinusoidal with angular frequency ω. However, the amplitudes and phases of these sinusoidal voltages and currents are all different. For example, the voltage across a resistor might lead the voltage across a capacitor by 90° ($\frac{\pi}{2}$ radians) and lag the voltage across an inductor by 90° ($\frac{\pi}{2}$ radians).

In order to make our application of phasors to electrical systems concrete, we consider the series RLC circuit illustrated in Figure 3.12. The arrow labeled $i(t)$ denotes a current that flows in response to the voltage applied, and the + and − on the voltage source indicate that the polarity of the applied voltage is positive on the top and negative on the bottom. Our convention is that current flows from positive to negative, in this case clockwise in the circuit.

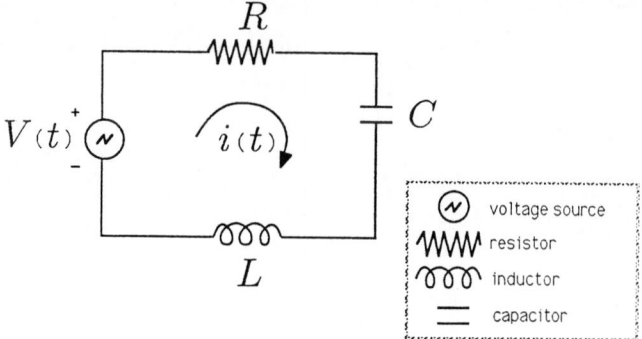

Figure 3.12: Series RLC Circuit

We will assume that the voltage source is an audio oscillator that produces the voltage

$$V(t) = A\cos(\omega t + \phi). \qquad 3.37$$

We represent this voltage as the complex signal

$$V(t) \quad \leftrightarrow \quad Ae^{j\phi}e^{j\omega t} \qquad 3.38$$

and give it the phasor representation

$$V(t) \quad \leftrightarrow \quad V; \qquad V = Ae^{j\phi}. \qquad 3.39$$

We then describe the voltage source by the phasor V and remember that we
can always compute the actual voltage by multiplying by $e^{j\omega t}$ and taking the
real part:

$$V(t) = \text{Re}\{Ve^{j\omega t}\}. \qquad 3.40$$

Problem 3.18   Show that $\text{Re}[Ve^{j\omega t}] = A\cos(\omega t + \phi)$ when $V = Ae^{j\phi}$. ∎

Circuit Laws. In your circuits classes you will study the Kirchhoff
laws that govern the low frequency behavior of circuits built from resistors
(R), inductors (L), and capacitors (C). In your study you will learn that the

voltage dropped across a resistor is related to the current that flows through it by the equation

$$V_R(t) = Ri(t). \qquad\qquad 3.41$$

You will learn that the voltage dropped across an inductor is proportional to the derivative of the current that flows through it, and the voltage dropped across a capacitor is proportional to the integral of the current that flows through it:

$$V_L(t) = L\frac{di}{dt}(t) \qquad\qquad 3.42$$

$$V_C(t) = \frac{1}{C}\int i(t)\,dt.$$

Phasors and Complex Impedance. Now suppose that the current in the preceding equations is sinusoidal, of the form

$$i(t) = B\cos(\omega t + \theta). \qquad\qquad 3.43$$

We may rewrite $i(t)$ as

$$i(t) = \text{Re}\{Ie^{j\omega t}\} \qquad\qquad 3.44$$

where I is the phasor representation of $i(t)$.

Problem 3.19 Find the phasor I in terms of B and θ in Equation 3.44. ∎
The voltage dropped across the resistor is

$$\begin{aligned} V_R(t) &= Ri(t) \\ &= R\,\text{Re}\{Ie^{j\omega t}\} \qquad\qquad 3.45 \\ &= \text{Re}\{RIe^{j\omega t}\}. \end{aligned}$$

Thus the phasor representation for $V_R(t)$ is

$$V_R(t) \quad\leftrightarrow\quad V_R; \qquad V_R = RI. \qquad\qquad 3.46$$

We call R the *impedance* of the resistor because R is the scale constant that relates the "phasor voltage V_R" to the "phasor current I."

The voltage dropped across the inductor is

$$V_L(t) = L\frac{di}{dt}(t) = L\frac{d}{dt}\,\text{Re}\{Ie^{j\omega t}\}. \qquad 3.47$$

The derivative may be moved through the Re[ ] operator (see Problem 3.20) to produce the result

$$V_L(t) = L\,\text{Re}\{j\omega Ie^{j\omega t}\}$$
$$= \text{Re}\{j\omega LIe^{j\omega t}\}. \qquad 3.48$$

Thus the phasor representation of $V_L(t)$ is

$$V_L(t) \quad \leftrightarrow \quad V_L; \qquad V_L = j\omega LI. \qquad 3.49$$

We call $j\omega L$ the impedance of the inductor because $j\omega L$ is the complex scale constant that relates "phasor voltage V_L" to "phasor current I."

Problem 3.20 Prove that the operators $\frac{d}{dt}$ and Re[ ] commute:

$$\frac{d}{dt}\,\text{Re}\{e^{j\omega t}\} = \text{Re}\left\{\frac{d}{dt}\,e^{j\omega t}\right\}. \quad \blacksquare$$

The voltage dropped across the capacitor is

$$V_C(t) = \frac{1}{C}\int i(t)\,dt = \frac{1}{C}\int \text{Re}\{Ie^{j\omega t}\}\,dt. \qquad 3.50$$

The integral may be moved through the Re[ ] operator to produce the result

$$V_C(t) = \frac{1}{C}\,\text{Re}\left\{\frac{I}{j\omega}\,e^{j\omega t}\right\}$$
$$= \text{Re}\left\{\frac{I}{j\omega C}\,e^{j\omega t}\right\}. \qquad 3.51$$

Thus the phasor representation of $V_C(t)$ is

$$V_C(t) \quad \leftrightarrow \quad V_C; \qquad V_C = \frac{I}{j\omega C}. \qquad 3.52$$

We call $\frac{1}{j\omega C}$ the *impedance* of the capacitor because $\frac{1}{j\omega C}$ is the complex scale constant that relates "phasor voltage V_C" to "phasor current I."

Kirchhoff's Voltage Law. Kirchhoff's voltage law says that the voltage dropped in the series combination of R, L, and C illustrated in Figure 3.12 equals the voltage generated by the source (this is one of two fundamental conservation laws in circuit theory, the other being a conservation law for current):

$$V(t) = V_R(t) + V_L(t) + V_C(t).$$ 3.53

If we replace all of these voltages by their complex representations, we have

$$\text{Re}\{Ve^{j\omega t}\} = \text{Re}\{(V_R + V_L + V_C)e^{j\omega t}\}.$$ 3.54

An obvious solution is

$$\begin{aligned} V &= V_R + V_L + V_C \\ &= \left(R + j\omega L + \frac{1}{j\omega C}\right)I \end{aligned}$$ 3.55

where I is the phasor representation for the current that flows in the circuit. This solution is illustrated in Figure 3.13, where the phasor voltages RI, $j\omega LI$, and $\frac{1}{j\omega C}I$ are forced to add up to the phasor voltage V.

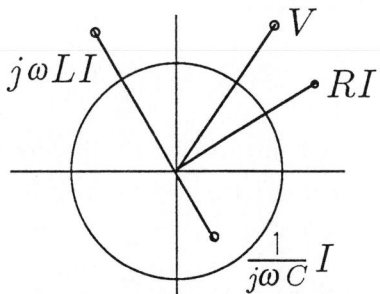

Figure 3.13: Phasor Addition to Satisfy Kirchhoff's Law

Problem 3.21 Redraw Figure 3.13 for $R = \omega L = \frac{1}{\omega C} = 1$. ∎

Impedance. We call the complex number $R + j\omega L + \frac{1}{j\omega C}$ the complex impedance for the series RLC network because it is the complex number that relates the phasor voltage V to the phasor current I:

$$V = ZI \qquad\qquad 3.56$$

$$Z = R + j\omega L + \frac{1}{j\omega C}.$$

The complex number Z depends on the numerical values of resistance (R), inductance (L), and capacitance (C), but it also depends on the angular frequency (ω) used for the sinusoidal source. This impedance may be manipulated as follows to put it into an illuminating form:

$$
\begin{aligned}
Z &= R + j\left(\omega L - \frac{1}{\omega C}\right) \\
&= R + j\sqrt{\frac{L}{C}}\left(\omega\sqrt{LC} - \frac{1}{\omega\sqrt{LC}}\right).
\end{aligned}
\qquad\qquad 3.57
$$

The parameter $\omega_0 = \frac{1}{\sqrt{LC}}$ is a parameter that you will learn to call an "undamped natural frequency" in your more advanced circuits courses. With it, we may write the impedance as

$$Z = R + j\omega_0 L\left(\frac{\omega}{\omega_0} - \frac{\omega_0}{\omega}\right). \qquad\qquad 3.58$$

The frequency $\frac{\omega}{\omega_0}$ is a normalized frequency that we denote by ν. Then the impedance, as a function of normalized frequency, is

$$Z(\nu) = R + j\omega_0 L\left(\nu - \frac{1}{\nu}\right). \qquad\qquad 3.59$$

When the normalized frequency equals one ($\nu = 1$), then the impedance is entirely real and $Z = R$. The circuit looks like it is a single resistor.

The magnitude and phase of the impedance $Z(\nu)$ are

$$|Z(\nu)| = R\left[1 + \left(\frac{\omega_0 L}{R}\right)^2 \left(\nu - \frac{1}{\nu}\right)^2\right]^{1/2} \qquad 3.60$$

$$\arg Z(\nu) = \tan^{-1} \frac{\omega_0 L}{R}\left(\nu - \frac{1}{\nu}\right). \qquad 3.61$$

The impedance obeys the following symmetries around $\nu = 1$:

$$Z(\nu) = Z^*\left(\frac{1}{\nu}\right) \qquad 3.62$$

$$|Z(\nu)| = \left|Z\left(\frac{1}{\nu}\right)\right|$$

$$\arg Z(\nu) = -\arg Z\left(\frac{1}{\nu}\right).$$

In the next paragraph we show how this impedance function influences the current that flows in the circuit.

Resonance. The phasor representation for the current that flows in the series *RLC* circuit is

$$
\begin{aligned}
I &= \frac{V}{Z(\nu)} \\
&= \frac{1}{|Z(\nu)|} e^{-j \arg Z(\nu)} V.
\end{aligned}
\qquad 3.63
$$

The function $H(\nu) = \frac{1}{Z(\nu)}$ displays a "resonance phenomenon." That is, $|H(\nu)|$ peaks at $\nu = 1$ and decreases to zero at $\nu = 0$ and $\nu = \infty$:

$$|H(\nu)| = \begin{cases} 0, & \nu = 0 \\ \frac{1}{R}, & \nu = 1 \\ 0, & \nu = \infty. \end{cases} \qquad 3.64$$

When $|H(\nu)| = 0$, no current flows.

The function $|H(\nu)|$ is plotted against the normalized frequency $\nu = \frac{\omega}{\omega_0}$ in Figure 3.14. The resonance peak occurs at $\nu = 1$, where $|H(\nu)| = \frac{1}{R}$, meaning that the circuit looks purely resistive. Resonance phenomena underlie the frequency selectivity of all electrical and mechanical networks.

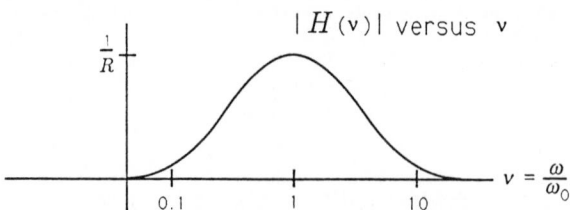

Figure 3.14: Resonance in a Series RLC Circuit

Problem 3.22 (MATLAB) Write a MATLAB program to compute and plot $|H(\nu)|$ and $\arg H(\nu)$ versus ν for ν ranging from 0.1 to 10 in steps of 0.1. Carry out your computations for $\omega_0 \frac{L}{R} = 10$, 1, 0.1, and 0.01, and overplot your results. ∎

Circle Criterion and Power Factor. Our study of the impedance $Z(\nu)$ and the function $H(\nu) = \frac{1}{Z(\nu)}$ brings insight into the resonance of an RLC circuit and illustrates the frequency selectivity of the circuit. But there is more that we can do to illuminate the behavior of the circuit.

Let's return to the equation for voltage conservation, Equation 3.55, and write it as

$$V = RI + j\left(\omega L - \frac{1}{\omega C}\right)I. \qquad 3.65$$

This equation shows how voltage is divided between resistor voltage RI and inductor-capacitor voltage $j\left(\omega L - \frac{1}{\omega C}\right)I$. If we use our definition of $\omega_0 = \frac{1}{\sqrt{LC}}$, we can rewrite this equation as

$$V = RI + j\omega_0 L\left(\frac{\omega}{\omega_0} - \frac{\omega_0}{\omega}\right)I$$

or

$$V = RI + \frac{j\omega_0 L}{R}\left(\nu - \frac{1}{\nu}\right)RI. \qquad 3.66$$

In order to simplify our notation, we can write this equation as

$$V = V_R + jk(\nu)V_R \qquad \text{3.67}$$

where V_R is the phasor voltage RI and $k(\nu)$ is the real variable

$$k(\nu) = \frac{\omega_0 L}{R}\left(\nu - \frac{1}{\nu}\right). \qquad \text{3.68}$$

Equation 3.67 brings very important geometrical insights. First, even though the phasor voltage V_R in the RLC circuit is complex, the terms V_R and $jk(\nu)V_R$ are out of phase by $\frac{\pi}{2}$ radians. This means that, for every allowable value of V_R, the corresponding $jk(\nu)V_R$ must add in a right triangle to produce the source voltage V. This is illustrated in Figure 3.15(a). As the frequency ν changes, then $k(\nu)$ changes, producing other values of V_R and $jk(\nu)V_R$ that sum to V. Several such solutions for V_R and $jk(\nu)V_R$ are illustrated in Figure 3.15(b). From the figure we gain the clear impression that the phasor voltage V_R lies on a circle of radius $\frac{V}{2}$, centered at $\frac{V}{2}$. Let's try this solution,

$$\begin{aligned} V_R &= \frac{V}{2} + \frac{V}{2}e^{j\psi} \\ &= \frac{V}{2}(1 + e^{j\psi}), \end{aligned} \qquad \text{3.69}$$

and explore its consequences. When this solution is substituted into Equation 3.67, the result is

$$V = \frac{V}{2}(1 + e^{j\psi}) + jk(\nu)\frac{V}{2}(1 + e^{j\psi}) \qquad \text{3.70}$$

or

$$2 = (1 + e^{j\psi})\left[1 + jk(\nu)\right]. \qquad \text{3.71}$$

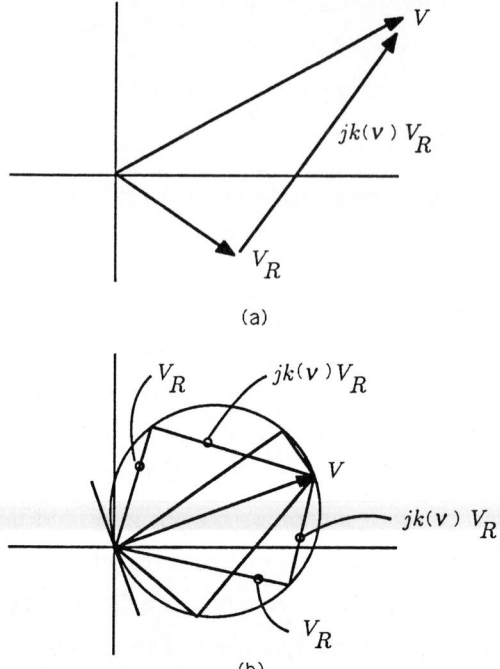

(a)

(b)

Figure 3.15: The Components of V; (a) Addition of V_R and $jk(\nu)V_R$ to Produce V, and (b) Several Values of V_R and $jk(\nu)V_R$ that Produce V

If we multiply the left-hand side by its complex conjugate and the right-hand side by its complex conjugate, we obtain the identity

$$4 = 2(1 + \cos\psi)\left[1 + k^2(\nu)\right]. \qquad 3.72$$

This equation tells us how the angle ψ depends on $k(\nu)$ and, conversely, how $k(\nu)$ depends on ψ:

$$\cos\psi = \frac{1 - k^2(\nu)}{1 + k^2(\nu)}$$

$$k^2(\nu) = \frac{1 - \cos\psi}{1 + \cos\psi}.$$

The number $\cos\psi$ lies between -1 and $+1$, so a circular solution does indeed work.

Problem 3.23 Check $-1 \leq \cos \psi \leq 1$ for $-\infty < k < \infty$ and $-\infty < k < \infty$ for $-\pi \leq \psi \leq \pi$. Sketch k versus ψ and ψ versus k. ∎

The equation $V_R = \frac{V}{2}(1 + e^{j\psi})$ is illustrated in Figure 3.16. The angle that V_R makes with V is determined from the equation

$$2\phi + \pi - \psi = \pi \implies \phi = \frac{\psi}{2}. \qquad 3.73$$

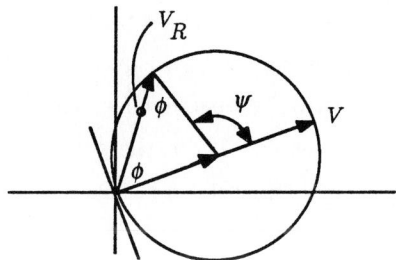

Figure 3.16: The Voltages V and V_R, and the Power Factor $\cos \phi$

In the study of power systems, $\cos \phi$ is a "power factor" that determines how much power is delivered to the resistor. We may denote the power factor as

$$\eta = \cos \phi = \cos \frac{\psi}{2}. \qquad 3.74$$

But $\cos \psi$ may be written as

$$\cos \psi = \cos(\phi + \phi) = \cos^2 \phi - \sin^2 \phi$$
$$= \cos^2 \phi - (1 - \cos^2 \phi)$$
$$= 2 \cos^2 \phi - 1 \qquad 3.75$$
$$= 2\eta^2 - 1.$$

Therefore the square of the power factor η is

$$\eta^2 = \frac{\cos \psi + 1}{2} = \frac{1}{1 + k^2(\nu)}. \qquad 3.76$$

The power factor is a maximum of 1 for $k(\nu) = 0$, corresponding to $\nu = 1$ ($\omega = \omega_0$). It is a minimum of 0 for $k(\nu) = \pm\infty$, corresponding to $\nu = 0, \infty$ ($\omega = 0, \infty$).

Problem 3.24 With k defined as $k(\nu) = \frac{\omega_0 L}{R}\left(\nu - \frac{1}{\nu}\right)$, plot $k^2(\nu)$, $\cos\psi$, and η^2 versus ν. ∎

Problem 3.25 Find the value of ν that makes the power factor $\eta = 0.707$. ∎

*3.7 Light Scattering by a Slit

One of the most spectacular successes for phasor analysis arises in the study of light diffraction by a narrow slit. The experiment is to shine laser light through a slit in an otherwise opaque sheet and observe the pattern of light that falls on a distant screen. When the slit is very narrow compared with the wavelength of the light, then the light that falls on the screen is nearly uniform in intensity. However, when the width of the slit is comparable to the wavelength of the light, the pattern of light that falls on the screen is scalloped in intensity, showing alternating light and dark bands. The experiment, and the observed results, are illustrated in Figure 3.17.

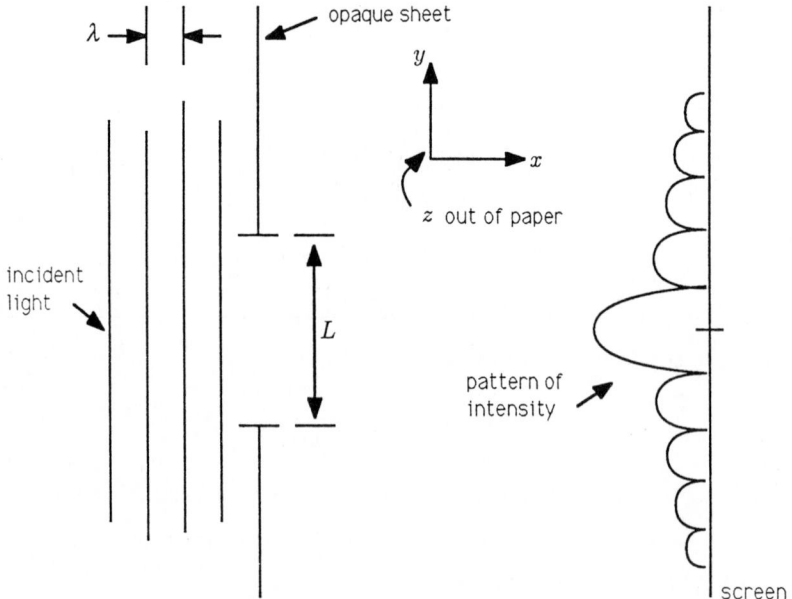

Figure 3.17: Light Diffraction by a Slit

Why should this experiment produce this result? Phasor analysis illuminates the question and produces an elegant mathematical description of a profoundly important optical experiment.

Huygens's Principle. We will assume, as Christiaan Huygens did, that the light incident on the slit sets up a light field in the slit that may be modeled by N discrete sources, each of which radiates a "spherical wave of light." This model is illustrated in Figure 3.18. The distance between sources is d, and $Nd = L$ is the width of the slit. Each source is indexed by n, and n runs from 0 to $N-1$. The 0th source is located at the origin of our coordinate system.

The spherical wave radiated by the nth source is described by the equation

$$E(r_n, t) = \text{Re}\left\{ \frac{A}{N} e^{j\left[\omega t - (2\pi/\lambda)r_n\right]} \right\}. \qquad 3.77$$

The function $E(r_n, t)$ describes the "electric field" at time t and distance r_n from the nth source. The field is constant as long as the variable $\omega t - (2\pi/\lambda)r_n$ is constant. Therefore, if we freeze time at $t = t_0$, the field will be constant on a sphere of radius r_n. This is illustrated in Figure 3.18.

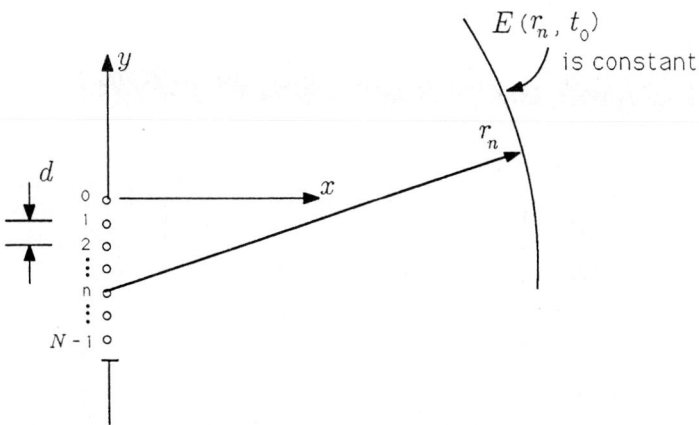

Figure 3.18: Huygens's Model for Light Diffraction

Problem 3.26   Fix r_n in $E(r_n, t)$ and show that $E(r_n, t)$ varies cosinusoidally with time t. Sketch the function and interpret it. What is its period? ∎

Problem 3.27   Fix t in $E(r_n, t)$ and show that $E(r_n, t)$ varies cosinusoidally with radius r_n. Sketch the function and interpret it. Call the "period in r_n" the "wavelength." Show that the wavelength is λ. ∎

Problem 3.28   The "crest of the wave $E(r_n, t)$" occurs when $\omega t - (2\pi/\lambda)r_n = 0$. Show that the crest moves through space at velocity $v = \omega\lambda/2\pi$. ∎

Geometry.   If we now pick a point P on a distant screen, that point will be at distance r_0 from source $0, \ldots, r_n$ from source $n, \ldots$, and so on. If we isolate the sources 0 and n, then we have the geometric picture of Figure 3.19. The angle ϕ is the angle that point P makes with the horizontal axis. The Pythagorean theorem says that the connection between distances r_0 and r_n is

$$(r_n - nd\sin\phi)^2 + (nd\cos\phi)^2 = r_0^2. \qquad\qquad 3.78$$

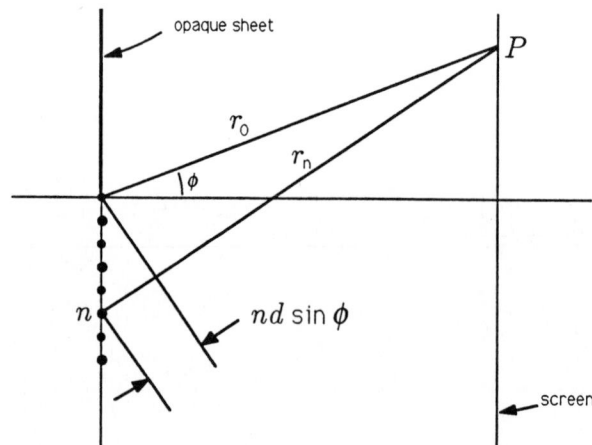

Figure 3.19: Geometry of the Experiment

Let's try the solution

$$r_n = r_0 + nd\sin\phi. \qquad\qquad 3.79$$

This solution produces the approximate identity

$$r_0^2 + (nd\cos\phi)^2 \cong r_0^2 \qquad\qquad 3.80$$

$$1 + \left(\frac{nd}{r_0}\cos\phi\right)^2 \cong 1.$$

This will be close for $\frac{nd}{r_0} \ll 1$. We will assume that the slit width L is small compared to the distance to any point on the screen. Then $\frac{Nd}{r_0} = \frac{L}{r_0} \ll 1$, in which case the approximate solution for r_n is valid for all n. This means that, for any point P on the distant screen, the light contributed by the nth source is approximately

$$E_n(\phi, t) = \text{Re}\left\{\frac{A}{N}\, e^{j[\omega t - (2\pi/\lambda)\,(r_0 + nd\sin\phi)]}\right\}$$

$$= \text{Re}\left\{\frac{A}{N}\, e^{-j(2\pi/\lambda)r_0}e^{-j(2\pi/\lambda)nd\sin\phi}e^{j\omega t}\right\}. \qquad\qquad 3.81$$

The phasor representation for this function is just

$$E_n(\phi) = \frac{A}{N}\, e^{-j(2\pi/\lambda)r_0}e^{-j(2\pi/\lambda)nd\sin\phi}. \qquad\qquad 3.82$$

Note that $E_0(\phi)$, the phasor associated with the 0th source, is $\frac{A}{N}\, e^{-j(2\pi/\lambda)r_0}$. Therefore we may write the phasor representation for the light contributed by the nth source to be

$$E_n(\phi) = E_0(\phi)e^{-j(2\pi/\lambda)nd\sin\phi}. \qquad\qquad 3.83$$

This result is very important because it shows the light arriving at point P from different sources to be "out of phase" by an amount that depends on the ratio $\frac{nd\sin\phi}{\lambda}$.

Phasors and Interference. The phasor representation for the field observed at point P on the screen is the sum of the phasors contributed by each source:

$$E(\phi) = \sum_{n=0}^{N-1} E_n(\phi) = E_0(\phi)\sum_{n=0}^{N-1} e^{-j(2\pi/\lambda)nd\sin\phi}. \qquad\qquad 3.84$$

This is a sum of the form

$$E(\phi) = E_0(\phi) \sum_{n=0}^{N-1} e^{jn\theta}$$

3.85

where the angle θ is $(2\pi/\lambda)\, d \sin\phi$. This sum is illustrated in Figure 3.20 for several representative values of θ. Note that for small θ, meaning small ϕ, the sum has large magnitude, whereas for θ on the order of $2\pi/N$, the sum is small. This simple geometric interpretation shows that for some values of ϕ, corresponding to some points P on the screen, there will be constructive interference between the phasors, while for other values of ϕ there will be destructive interference. Constructive interference produces bright light, and destructive interference produces darkness.

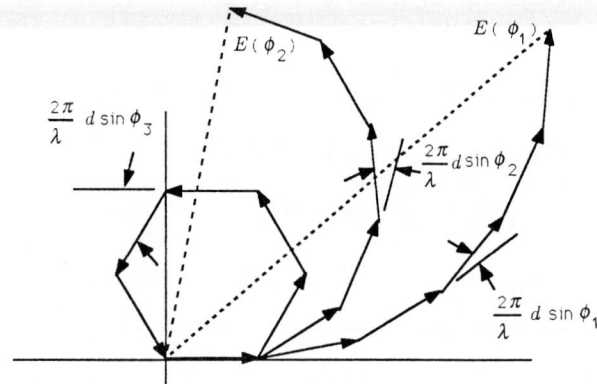

Figure 3.20: Phasor Sums for Diffraction

The geometry of Figure 3.20 is illuminating. However, we already know from our study of complex numbers and geometric sums that the phasor sum of Equation 3.85 may be written as

$$E(\phi) = E_0(\phi)\frac{1 - e^{-j(2\pi/\lambda)Nd\sin\phi}}{1 - e^{-j(2\pi/\lambda)\,d\sin\phi}}.$$

3.86

This result may be manipulated to produce the form

$$E(\phi) = \frac{A}{N} e^{-j(2\pi/\lambda)r_0} e^{-j(\pi/\lambda)(N-1)d\sin\phi} \frac{\sin\left(N\pi \frac{d}{\lambda}\sin\phi\right)}{\sin\left(\pi \frac{d}{\lambda}\sin\phi\right)}.$$

3.87

The magnitude is the intensity of the light at angle ϕ from horizontal:

$$|E(\phi)| = \left| \frac{A}{N} \frac{\sin\left(N\pi \frac{d}{\lambda} \sin\phi\right)}{\sin\left(\pi \frac{d}{\lambda} \sin\phi\right)} \right|. \qquad 3.88$$

Problem 3.29 Derive Equation 3.87 from Equation 3.86. ∎

Limiting Form. Huygens's model is exact when d shrinks to 0 and N increases to infinity in such a way that $Nd \to L$, the slit width. Then

$$|E(\phi)| \;\longrightarrow\; \left| \frac{A\sin\left(\frac{\pi L}{\lambda} \sin\phi\right)}{\frac{\pi L}{\lambda} \sin\phi} \right|. \qquad 3.89$$

This function is plotted in Figure 3.21 for two values of $\frac{L}{\lambda}$, the width of the slit measured in wavelengths. When $\frac{L}{\lambda} << 1$ (i.e., $\frac{\lambda}{L} >> 1$), then the light is uniformly distributed on the screen. However, when $\frac{L}{\lambda} > 1$ $\left(\frac{\lambda}{L} < 1\right)$, then the function has many zeros for $|\sin\phi| < 1$, as illustrated in the figure. These zeros correspond to dark spots on the screen where the fields radiated from the infinity of points within the slit interfere destructively.

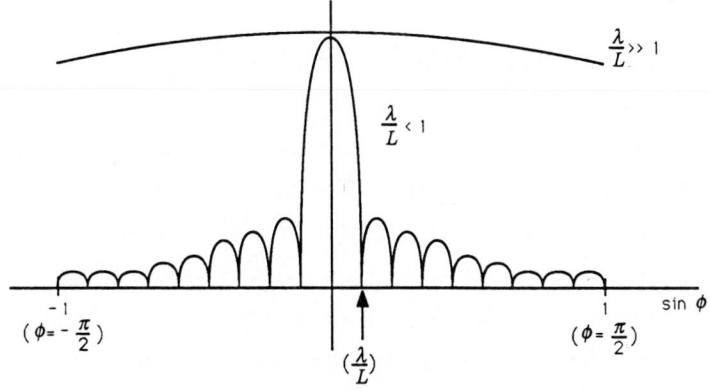

Figure 3.21: Interference Pattern for a Slit

Problem 3.30 Derive Equation 3.89 from Equation 3.88. ∎

Problem 3.31 (MATLAB) Plot the discrete approximation

$$\left| \frac{\frac{A}{N} \sin\left(N\pi \frac{d}{\lambda} \sin\phi\right)}{\sin\left(\pi \frac{d}{\lambda} \sin\phi\right)} \right|$$

versus $\sin\phi$ for $\frac{L}{\lambda} = \frac{Nd}{\lambda} = 10$ and $N = 2, 4, 8, 16, 32$. Compare with the continuous, limiting form

$$\left| \frac{A \sin\left(\pi \frac{L}{\lambda} \sin\phi\right)}{\pi \frac{L}{\lambda} \sin\phi} \right|. \quad ∎$$

3.8 Numerical Experiment (Interference Patterns)

Let's add two sinusoids whose amplitudes and frequencies are identical and whose phases are different:

$$x(t) = A\cos(\omega t + \phi) + A\cos(\omega t + \phi + \psi).$$

Show analytically that this sum has the phasor representation

$$X = 2A\cos\left(\frac{\psi}{2}\right) e^{j[\phi + (\psi/2)]}.$$

Interpret this finding. Then write a MATLAB program that computes and plots complex X on the complex plane as ψ varies from 0 to 2π and that plots magnitude, $|X|$, and phase, $\arg X$, versus the phase angle ψ. (You will have to choose $\psi = n\frac{2\pi}{N}$, $n = 0, 1, \ldots, N-1$, for a suitable N.) When do you get constructive interference and when do you get destructive interference? Now compute and plot $x(t)$ versus t (you will need to discretize t) for several interesting values of ψ. Explain your interference results in terms of the amplitude and phase of $x(t)$ and the magnitude and phase of X. Use the subplots discussed in Appendix 1 to plot all of your results together.

| | | | | | | | | | | **4** |

Linear Algebra

4.9   Circuit Analysis

4.10   Numerical Experiment (Circuit Design)

Notes to Teachers and Students:

We use this chapter to introduce students to the algebraic structure of *vectors* and *matrices* and to introduce them to matrix computations. These matrix computations are used in Chapters 5 through 7 to solve problems in vector graphics, filtering, and binary coding.

Vectors are introduced in Section 4.2, along with algebraic and geometric interpretations of some fundamental vector operations and properties. Sections 4.3 through 4.5 introduce inner products and their applications, including norm, direction cosines, orthogonality, and projections. Some important alternatives to the Euclidean norm are introduced in Section 4.6. Matrices are motivated and introduced in Section 4.7. The notation in these sections can be daunting to the beginner, so we proceed very carefully, using example after example. In Section 4.8 we codify the elimination procedures that students have used in high school to solve linear systems of equations. The MATLAB demonstration in Demo 4.2 shows how to use MATLAB to solve linear equations. Section 4.9 shows how linear algebra and MATLAB can be used to analyze dc circuits. The numerical experiment in Section 4.10 gives students practice in building function files in MATLAB and shows how to solve a sequence of linear equations in order to *design* a circuit with desired properties.

Occasionally we have placed important results in the problems. We feel that students should not miss the material in Problems 4.3, 4.6, 4.20, 4.22, and 4.25.

4.1   Introduction

Linear algebra is a branch of mathematics that is used by engineers and applied scientists to design and analyze complex systems. Civil engineers use linear algebra to design and analyze load-bearing structures such as bridges. Mechanical engineers use linear algebra to design and analyze suspension systems, and electrical engineers use it to design and analyze electrical circuits. Electrical, biomedical, and aerospace engineers use linear algebra to enhance X rays, tomographs, and images from space. In this chapter and the next we study two common problems from electrical engineering and use linear algebra to solve them. The two problems are (i) electrical circuit analysis and (ii) coordinate transformations for computer graphics. The first of these applications requires us to understand the solution of linear systems of equations, and the second requires us to understand the representation of mathematical operators with matrices.

Much of linear algebra is concerned with systematic techniques for organizing and solving simultaneous linear equations by elimination and substitution. The following example illustrates the basic ideas that we intend to develop.

Example 4.1. A woman steps onto a moving sidewalk at a large airport and stands while the moving sidewalk moves her forward at 1.2 meters/second. At the same time, a man begins walking against the motion of the sidewalk from the opposite end at 1.5 meters/second (relative to the sidewalk). If the moving sidewalk is 85 meters long, how far does each person travel (relative to the ground) before they pass each other?

To solve this problem, we first assign a variable to each unknown quantity. Let x_1 be the distance traveled by the woman, and let x_2 be the distance traveled by the man. The sum of the two distances is 85 meters, giving us one equation:

$$x_1 + x_2 = 85. \qquad 4.1$$

Our second equation is based on the time required before they pass. Time equals distance divided by rate, and the time is the same for both people:

$$\frac{x_1}{1.2} = \frac{x_2}{1.5 - 1.2} \quad \Longrightarrow \quad 0.3x_1 - 1.2x_2 = 0. \qquad 4.2$$

We may substitute Equation 4.2 into Equation 4.1 to obtain the result $\frac{1.2}{0.3} x_2 +$ $x_2 = 85$, or

$$5x_2 = 85 \implies x_2 = 17. \qquad\qquad 4.3$$

Combining the result from Equation 4.3 with that of Equation 4.1, we find that

$$x_1 = 68. \qquad\qquad 4.4$$

So the man travels 17 meters, and the woman travels 68 meters. ☐

Equations 4.1 and 4.2 are the key equations of Example 4.1. They may be organized into the "matrix equation"

$$\begin{bmatrix} 1 & 1 \\ 0.3 & -1.2 \end{bmatrix} \begin{bmatrix} x_1 \\ x_2 \end{bmatrix} = \begin{bmatrix} 85 \\ 0 \end{bmatrix}. \qquad\qquad 4.5$$

The rules for matrix-vector multiplication are evidently

$$(1)x_1 + (1)x_2 = 85$$

$$(0.3)x_1 + (-1.2)x_2 = 0.$$

Equations 4.2 and 4.3 may be organized into the matrix equation

$$\begin{bmatrix} 0 & 5 \\ 0.3 & -1.2 \end{bmatrix} \begin{bmatrix} x_1 \\ x_2 \end{bmatrix} = \begin{bmatrix} 85 \\ 0 \end{bmatrix}. \qquad\qquad 4.6$$

This equation represents one partially solved form of Equation 4.5, wherein we have used the so-called *Gauss elimination procedure* to introduce a zero into the matrix equation in order to isolate one variable. The MATLAB software contains built-in procedures to implement Gauss elimination on much larger matrices. Thus MATLAB may be used to solve large systems of linear equations.

Before we can apply linear algebra to more interesting physical problems, we need to introduce the mathematical tools we will use.

4.2 Vectors

For our purposes, a *vector* is a collection of *real numbers* in a one-dimensional array.[1] We usually think of the array as being arranged in a column and write

$$
\mathbf{x} = \begin{bmatrix} x_1 \\ x_2 \\ x_3 \\ \vdots \\ x_n \end{bmatrix}.
\tag{4.7}
$$

Notice that we indicate a vector with boldface and the constituent elements with subscripts. A real number by itself is called a *scalar*, in distinction from a vector or a matrix. We say that $\mathbf{x}$ is an n-vector, meaning that $\mathbf{x}$ has n elements. To indicate that x_1 is a real number, we write

$$
x_1 \in \mathcal{R},
\tag{4.8}
$$

meaning that x_1 is contained in $\mathcal{R}$, the set of real numbers. To indicate that $\mathbf{x}$ is a vector of n real numbers, we write

$$
\mathbf{x} \in \mathcal{R}^n,
\tag{4.9}
$$

meaning that $\mathbf{x}$ is contained in $\mathcal{R}^n$, the set of real n-tuples. Geometrically, $\mathcal{R}^n$ is n-dimensional space, and the notation $\mathbf{x} \in \mathcal{R}^n$ means that $\mathbf{x}$ is a point in that space, specified by the n coordinates $x_1, x_2, \ldots, x_n$. Figure 4.1 shows a vector in $\mathcal{R}^3$, drawn as an arrow from the origin to the point $\mathbf{x}$. Our geometric intuition begins to fail above three dimensions, but the linear algebra is completely general.

---

[1] In a formal development of linear algebra, the abstract concept of a vector space plays a fundamental role. We will leave such concepts to a complete course in linear algebra and introduce only the functional techniques necessary to solve the problems at hand.

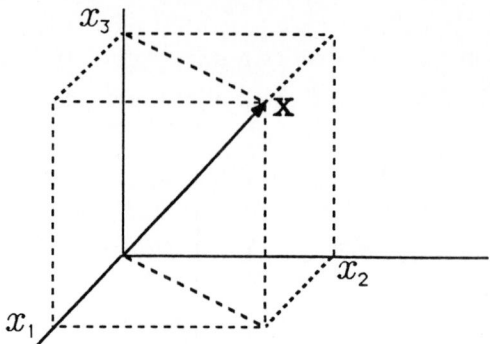

Figure 4.1: A Vector in $\mathcal{R}^3$

We sometimes find it useful to sketch vectors with more than three dimensions in the same way as the three-dimensional vector of Figure 4.1. We then consider each axis to represent more than one dimension, a hyperplane, in our n-dimensional space. We cannot show all the details of what is happening in n-space on a three-dimensional figure, but we can often show important features and gain geometrical insight.

Vector Addition. Vectors with the same number of elements can be added and subtracted in a very natural way:

$$
\mathbf{x} + \mathbf{y} = \begin{bmatrix} x_1 + y_1 \\ x_2 + y_2 \\ x_3 + y_3 \\ \vdots \\ x_n + y_n \end{bmatrix} \; ; \qquad \mathbf{x} - \mathbf{y} = \begin{bmatrix} x_1 - y_1 \\ x_2 - y_2 \\ x_3 - y_3 \\ \vdots \\ x_n - y_n \end{bmatrix} . \qquad 4.10
$$

Example 4.2. The difference between the vector $\mathbf{x} = \begin{bmatrix} 1 \\ 1 \\ 1 \end{bmatrix}$ and the

vector $\mathbf{y} = \begin{bmatrix} 0 \\ 0 \\ 1 \end{bmatrix}$ is the vector $\mathbf{z} = \mathbf{x} - \mathbf{y} = \begin{bmatrix} 1 \\ 1 \\ 0 \end{bmatrix}$. These vectors are illustrated in Figure 4.2. You can see that this result is consistent with the definition of vector subtraction in Equation 4.10. You can also picture the subtraction in Figure 4.2 by mentally reversing the direction of vector $\mathbf{y}$ to get $-\mathbf{y}$ and then adding it to $\mathbf{x}$ by sliding it to the position where its tail coincides with the head of vector $\mathbf{x}$. (The head is the end with the arrow.) When you slide a vector to a new position for adding to another vector, you must not change its length or direction. $\square$

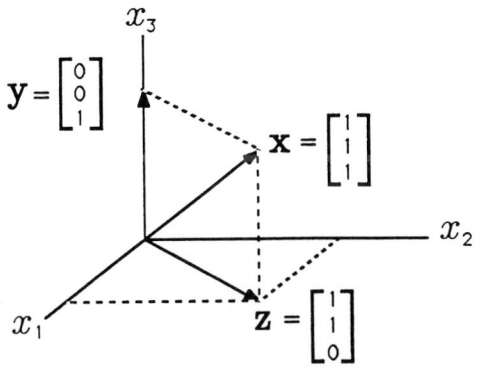

Figure 4.2: Subtraction of Vectors

Problem 4.1 Compute and plot $\mathbf{x} + \mathbf{y}$ and $\mathbf{x} - \mathbf{y}$ for each of the following cases:

(a) $\mathbf{x} = \begin{bmatrix} 1 \\ 3 \\ 2 \end{bmatrix}$, $\mathbf{y} = \begin{bmatrix} 1 \\ 2 \\ 3 \end{bmatrix}$;

(b) $\mathbf{x} = \begin{bmatrix} -1 \\ 3 \\ -2 \end{bmatrix}$, $\mathbf{y} = \begin{bmatrix} 1 \\ 2 \\ 3 \end{bmatrix}$;

(c) $\mathbf{x} = \begin{bmatrix} 1 \\ -3 \\ 2 \end{bmatrix}$, $\mathbf{y} = \begin{bmatrix} 1 \\ 3 \\ 2 \end{bmatrix}$. $\blacksquare$

Scalar Product. Several different kinds of vector multiplication are defined.[2] We begin with the *scalar product.* Scalar multiplication is defined for scalar a and vector $\mathbf{x}$ as

$$a\mathbf{x} = \begin{bmatrix} ax_1 \\ ax_2 \\ ax_3 \\ \vdots \\ ax_n \end{bmatrix}.$$ 4.11

If $|a| < 1$, then the vector $a\mathbf{x}$ is "shorter" than the vector $\mathbf{x}$; if $|a| > 1$, then the vector $a\mathbf{x}$ is "longer" than $\mathbf{x}$. This is illustrated for a 2-vector in Figure 4.3.

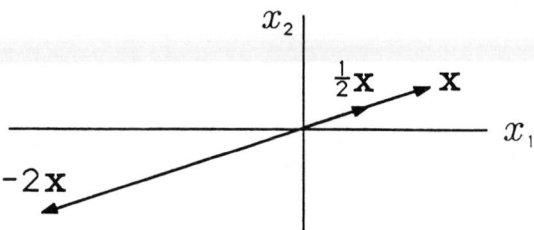

Figure 4.3: The Scalar Product $a\mathbf{x}$

Problem 4.2 Compute and plot the scalar product $a\mathbf{x}$ when $\mathbf{x} = \begin{bmatrix} 1 \\ 1/2 \\ 1/4 \end{bmatrix}$ for each of the following scalars:
  (a) $a = 1$;
  (b) $a = -1$;
  (c) $a = -1/4$;
  (d) $a = 2$. ∎

Problem 4.3 Given vectors $\mathbf{x}, \mathbf{y}, \mathbf{z} \in \mathcal{R}^n$ and the scalar $a \in \mathcal{R}$, prove the following identities:

---

[2] The division of two vectors is undefined, although three different "divisions" are defined in MATLAB.

(a) $\mathbf{x} + \mathbf{y} = \mathbf{y} + \mathbf{x}$. Is vector addition commutative?

(b) $(\mathbf{x} + \mathbf{y}) + \mathbf{z} = \mathbf{x} + (\mathbf{y} + \mathbf{z})$. Is vector addition associative?

(c) $a(\mathbf{x} + \mathbf{y}) = a\mathbf{x} + a\mathbf{y}$. Is scalar multiplication distributive over vector addition? ∎

4.3 Inner Product and Euclidean Norm

The inner product $(\mathbf{x}, \mathbf{y})$ between vectors $\mathbf{x}$ and $\mathbf{y}$ is a scalar consisting of the following sum of products:

$$(\mathbf{x}, \mathbf{y}) = x_1 y_1 + x_2 y_2 + x_3 y_3 + \cdots + x_n y_n. \qquad 4.12$$

This definition seems so arbitrary that we wonder what uses it could possibly have. We will show that the inner product has three main uses:

(i) computing length or "norm",

(ii) finding angles between vectors and checking for "orthogonality", and

(iii) computing the "component of one vector along another" (projection).

Since the inner product is so useful, we need to know what algebraic operations are permitted when we are working with inner products. The following are some properties of the inner product. Given $\mathbf{x}, \mathbf{y}, \mathbf{z} \in \mathcal{R}^n$ and $a \in \mathcal{R}$,

(i) $(\mathbf{x}, \mathbf{y}) = (\mathbf{y}, \mathbf{x})$;

(ii) $(a\mathbf{x}, \mathbf{y}) = a(\mathbf{x}, \mathbf{y}) = (\mathbf{x}, a\mathbf{y})$; and

(iii) $(\mathbf{x}, \mathbf{y} + \mathbf{z}) = (\mathbf{x}, \mathbf{y}) + (\mathbf{x}, \mathbf{z})$.

Problem 4.4 Prove the three preceding properties by using the definition of inner product. Is the equation $\mathbf{x}(\mathbf{y}, \mathbf{z}) = (\mathbf{x}, \mathbf{y})\mathbf{z}$ also a true property? Prove or give a counterexample. ∎

Euclidean Norm. Sometimes we want to measure the length of a vector, namely, the distance from the origin to the point specified by the vector's coordinates. A vector's length is called the *norm* of the vector. Recall from Euclidean geometry that the distance between two points is the square root of the sum of the squares of the distances in each dimension. Since we are measuring from the origin, this implies that the norm of the vector $\mathbf{x}$ is

$$\|\mathbf{x}\| = \sqrt{x_1^2 + x_2^2 + \cdots + x_n^2}.\tag{4.13}$$

Notice the use of the double vertical bars to indicate the norm. An equivalent definition of the norm, and of the norm squared, can be given in terms of the inner product:

$$\|\mathbf{x}\| = \sqrt{(\mathbf{x}, \mathbf{x})}\tag{4.14}$$

or

$$\|\mathbf{x}\|^2 = (\mathbf{x}, \mathbf{x}).$$

Example 4.3. The Euclidean norm of the vector

$$\mathbf{x} = \begin{bmatrix} 1 \\ 3 \\ 5 \\ -2 \end{bmatrix},\tag{4.15}$$

is $\|\mathbf{x}\| = \sqrt{1^2 + 3^2 + 5^2 + (-2)^2} = \sqrt{39}$. □

An important property of the norm and scalar product is that, for any $\mathbf{x} \in \mathcal{R}^n$ and $a \in \mathcal{R}$,

$$\|a\mathbf{x}\| = |a|\,\|\mathbf{x}\|.\tag{4.16}$$

So we can take a scalar multiplier outside of the norm if we take its absolute value.

Problem 4.5 Prove Equation 4.16. ∎

Cauchy-Schwarz Inequality. Inequalities can be useful engineering tools. They can often be used to find the best possible performance of a system, thereby telling you when to quit trying to make improvements (or proving to your boss that it can't be done any better). The most fundamental inequality in linear algebra is the Cauchy-Schwarz inequality. This inequality says that the inner product between two vectors $\mathbf{x}$ and $\mathbf{y}$ is less than or equal (in absolute value) to the norm of $\mathbf{x}$ times the norm of $\mathbf{y}$, with equality if and only if $\mathbf{y} = \alpha\mathbf{x}$:

$$|(\mathbf{x}, \mathbf{y})| \leq ||\mathbf{x}||\,||\mathbf{y}||. \qquad 4.17$$

To prove this theorem, we construct a third vector $\mathbf{z} = \lambda\mathbf{x} - \mathbf{y}$ and measure its norm squared:

$$||\lambda\mathbf{x} - \mathbf{y}||^2 = (\lambda\mathbf{x} - \mathbf{y}, \lambda\mathbf{x} - \mathbf{y}) = \lambda^2\,||\mathbf{x}||^2 - 2\lambda(\mathbf{x}, \mathbf{y}) + ||\mathbf{y}||^2 \geq 0. \qquad 4.18$$

So we have a polynomial in λ that is always greater than or equal to 0 (because every norm squared is greater than or equal to 0). Let's assume that $\mathbf{x}$ and $\mathbf{y}$ are given and minimize this norm squared with respect to λ. To do this, we take the derivative with respect to λ and equate it to 0:

$$2\lambda\,||\mathbf{x}||^2 - 2(\mathbf{x}, \mathbf{y}) = 0 \implies \lambda = \frac{(\mathbf{x}, \mathbf{y})}{||\mathbf{x}||^2}. \qquad 4.19$$

When this solution is substituted into the formula for the norm squared in Equation 4.18, we obtain

$$\left[\frac{(\mathbf{x}, \mathbf{y})}{||\mathbf{x}||^2}\right]^2 ||\mathbf{x}||^2 - \frac{2(\mathbf{x}, \mathbf{y})}{||\mathbf{x}||^2}(\mathbf{x}, \mathbf{y}) + ||\mathbf{y}||^2 \geq 0,$$

which simplifies to

$$-\frac{(\mathbf{x}, \mathbf{y})^2}{||\mathbf{x}||^2} + ||\mathbf{y}||^2 \geq 0 \implies (\mathbf{x}, \mathbf{y})^2 \leq ||\mathbf{x}||^2\,||\mathbf{y}||^2. \qquad 4.20$$

The proof of the Cauchy-Schwarz inequality is completed by taking the positive square root on both sides of Equation 4.20. When $\mathbf{y} = \alpha\mathbf{x}$, then

$$\begin{aligned}(\mathbf{x}, \mathbf{y})^2 = (\mathbf{x}, \alpha\mathbf{x})^2 &= \left[|\alpha|\,(\mathbf{x}, \mathbf{x})\right]^2 = \left(|\alpha|\,||\mathbf{x}||^2\right)^2 \\ &= \left(|\alpha|^2\,||\mathbf{x}||^2\right)||\mathbf{x}||^2 \\ &= (\alpha\mathbf{x}, \alpha\mathbf{x})\,||\mathbf{x}||^2 \qquad 4.21 \\ &= (\mathbf{y}, \mathbf{y})\,||\mathbf{x}||^2 \\ &= ||\mathbf{y}||^2\,||\mathbf{x}||^2,\end{aligned}$$

which shows that equality holds in Equation 4.17 when **y** is a scalar multiple of **x**.

Problem 4.6   Use the Cauchy-Schwarz inequality to prove the triangle inequality, which states

$$||\mathbf{x} + \mathbf{y}|| \leq ||\mathbf{x}|| + ||\mathbf{y}||.$$

Explain why this is called the *triangle inequality*. ∎

4.4 Direction Cosines

Unit Vectors. Corresponding to every vector **x** is a *unit vector* $\mathbf{u_x}$ pointing in the same direction as **x**. The term *unit vector* means that the norm of the vector is 1:

$$||\mathbf{u_x}|| = 1.$$

The question is, given **x**, how can we find $\mathbf{u_x}$? The first part of the answer is that $\mathbf{u_x}$ will have to be a positive scalar multiple of **x** in order to point in the same direction as **x**, as shown in Figure 4.4. Thus

$$\mathbf{u_x} = \alpha\mathbf{x}. \tag{4.22}$$

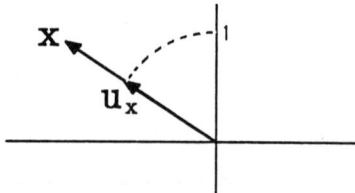

Figure 4.4: The Unit Vector $\mathbf{u_x}$

But what is α? We need to choose α so that the norm of $\mathbf{u_x}$ will be 1:

$$||\mathbf{u_x}|| = 1$$

$$||\alpha\mathbf{x}|| = 1$$

$$|\alpha| \, ||\mathbf{x}|| = 1$$

$$\alpha = \frac{1}{||\mathbf{x}||}. \qquad\qquad 4.23$$

We have dropped the absolute value bars on α because $||\mathbf{x}||$ is positive. The α that does the job is 1 over the norm of $\mathbf{x}$. Now we can write formulas for $\mathbf{u_x}$ in terms of $\mathbf{x}$ and $\mathbf{x}$ in terms of $\mathbf{u_x}$:

$$\mathbf{u_x} = \frac{1}{||\mathbf{x}||} \, \mathbf{x},$$

$$\mathbf{x} = ||\mathbf{x}|| \mathbf{u_x}. \qquad\qquad 4.24$$

So the vector $\mathbf{x}$ is its direction vector $\mathbf{u_x}$, scaled by its Euclidean norm.

Unit Coordinate Vectors. There is a special set of unit vectors called the unit coordinate vectors. The *unit coordinate vector* $\mathbf{e}_i$ is a unit vector in ∇^n that points in the positive direction of the ith coordinate axis. Figure 4.5 shows the three unit coordinate vectors in ∇^3.

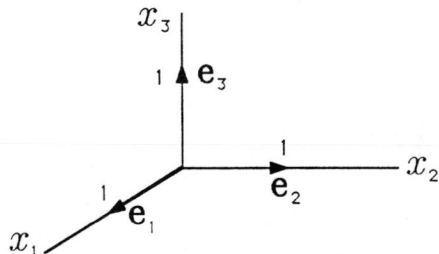

Figure 4.5: Unit Coordinate Vectors in $\mathcal{R}^3$

For three-dimensional space, $\mathcal{R}^3$, the unit coordinate vectors are

$$\mathbf{e}_1 = \begin{bmatrix} 1 \\ 0 \\ 0 \end{bmatrix}, \quad \mathbf{e}_2 = \begin{bmatrix} 0 \\ 1 \\ 0 \end{bmatrix}, \quad \mathbf{e}_3 = \begin{bmatrix} 0 \\ 0 \\ 1 \end{bmatrix}.$$

More generally, in n-dimensional space, there are n unit coordinate vectors given by

$$\mathbf{e}_i = \begin{bmatrix} 0 \\ \vdots \\ 0 \\ 1 \\ 0 \\ \vdots \\ 0 \end{bmatrix} \begin{matrix} \\ \\ \\ \longleftarrow \ i^{\text{th}} \text{ element} \\ \\ \\ \longleftarrow \ n \text{ elements.} \end{matrix} \qquad\qquad 4.25$$

You should satisfy yourself that this definition results in vectors with a norm of 1.

Problem 4.7 Find the norm of the vector $a\mathbf{u}$ where $\mathbf{u}$ is a unit vector. ∎

Problem 4.8 Find the unit vector $\mathbf{u_x}$ in the direction of

(a) $\mathbf{x} = \begin{bmatrix} 3 \\ 4 \end{bmatrix}$;

(b) $\mathbf{x} = \begin{bmatrix} 3 \\ 6 \\ -2 \end{bmatrix}$;

(c) $\mathbf{x} = \begin{bmatrix} 1 \\ -1 \\ 1 \\ -1 \end{bmatrix}$. ∎

Direction Cosines. We often say that vectors "have magnitude and direction." This is more or less obvious from Figure 4.1, where the three-dimensional vector $\mathbf{x}$ has length $\sqrt{x_1^2 + x_2^2 + x_3^2}$ and points in a direction defined by the components x_1, x_2, and x_3. It is perfectly obvious from Equation 4.24 where $\mathbf{x}$ is written as $\mathbf{x} = \|\mathbf{x}\|\mathbf{u_x}$. But perhaps there is another representation for a vector that places the notion of magnitude and direction in even clearer evidence.

Figure 4.6 shows an arbitrary vector $\mathbf{x} \in \mathcal{R}^3$ and the three-dimensional *unit coordinate vectors* $\mathbf{e}_1, \mathbf{e}_2, \mathbf{e}_3$. The inner product between the vector $\mathbf{x}$ and the unit vector $\mathbf{e}_k$ just reads out the kth component of $\mathbf{x}$:

$$(\mathbf{x}, \mathbf{e}_k) = (\mathbf{e}_k, \mathbf{x}) = x_k. \qquad 4.26$$

Since this is true even in $\mathcal{R}^n$, any vector $\mathbf{x} \in \mathcal{R}^n$ has the following representation in terms of unit vectors:

$$\mathbf{x} = (\mathbf{x}, \mathbf{e}_1)\mathbf{e}_1 + (\mathbf{x}, \mathbf{e}_2)\mathbf{e}_2 + \cdots + (\mathbf{x}, \mathbf{e}_n)\mathbf{e}_n. \qquad 4.27$$

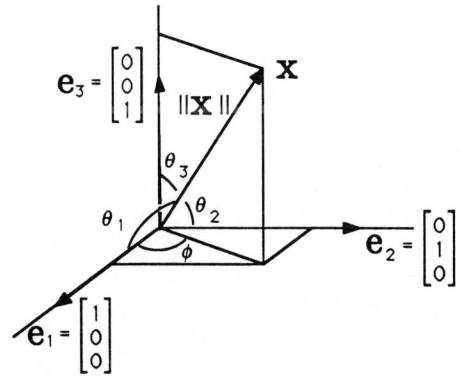

Figure 4.6: The Three-Dimensional Unit Vectors

Let us now generalize our notion of an angle θ between two vectors to $\mathcal{R}^n$ as follows:

$$\cos \theta = \frac{(\mathbf{x}, \mathbf{y})}{\|\mathbf{x}\| \|\mathbf{y}\|}. \qquad 4.28$$

The celebrated Cauchy-Schwarz inequality establishes that $-1 \leq \cos \theta \leq 1$. With this definition of angle, we may define the angle θ_k that a vector makes with the unit vector $\mathbf{e}_k$ to be

$$\cos \theta_k = \frac{(\mathbf{x}, \mathbf{e}_k)}{\|\mathbf{x}\| \|\mathbf{e}_k\|}. \qquad 4.29$$

But the norm of $\mathbf{e}_k$ is 1, so

$$\cos \theta_k = \frac{(\mathbf{x}, \mathbf{e}_k)}{\|\mathbf{x}\|} = \frac{x_k}{\|\mathbf{x}\|}. \qquad 4.30$$

When this result is substituted into the representation of $\mathbf{x}$ in Equation 4.27, we obtain the formula

$$\mathbf{x} = \|\mathbf{x}\|\cos\theta_1\mathbf{e}_1 + \|\mathbf{x}\|\cos\theta_2\mathbf{e}_2 + \cdots + \|\mathbf{x}\|\cos\theta_n\mathbf{e}_n$$
$$= \|\mathbf{x}\|(\cos\theta_1\mathbf{e}_1 + \cos\theta_2\mathbf{e}_2 + \cdots + \cos\theta_n\mathbf{e}_n). \qquad 4.31$$

This formula really shows that the vector $\mathbf{x}$ has "magnitude" $\|\mathbf{x}\|$ and direction $(\theta_1,\theta_2,\ldots,\theta_n)$ and that the magnitude and direction are sufficient to determine $\mathbf{x}$. We call $(\cos\theta_1,\cos\theta_2,\ldots,\cos\theta_n)$ the *direction cosines* for the vector $\mathbf{x}$. In the three-dimensional case, they are illustrated in Figure 4.6.

Problem 4.9   Show that Equation 4.28 agrees with the usual definition of an angle in two dimensions. ∎

Problem 4.10   Find the cosine of the angle between $\mathbf{x}$ and $\mathbf{y}$ where

(a) $\mathbf{x} = \begin{bmatrix} 1 \\ 0 \\ 0 \end{bmatrix}$, $\mathbf{y} = \begin{bmatrix} 2 \\ 2 \\ 2 \end{bmatrix}$;

(b) $\mathbf{x} = \begin{bmatrix} 1 \\ -1 \\ 1 \\ -1 \end{bmatrix}$, $\mathbf{y} = \begin{bmatrix} -1 \\ 0 \\ 1 \\ 0 \end{bmatrix}$;

(c) $\mathbf{x} = \begin{bmatrix} 2 \\ 1 \\ -2 \end{bmatrix}$, $\mathbf{y} = \begin{bmatrix} 4 \\ 2 \\ -4 \end{bmatrix}$. ∎

If we compare Equations 4.24 and 4.31 we see that the direction vector $\mathbf{u_x}$ is composed of direction cosines:

$$\mathbf{u_x} = \cos\theta_1\mathbf{e}_1 + \cos\theta_2\mathbf{e}_2 + \cdots + \cos\theta_n\mathbf{e}_n = \begin{bmatrix} \cos\theta_1 \\ \cos\theta_2 \\ \vdots \\ \cos\theta_n \end{bmatrix}. \qquad 4.32$$

With this definition we can write Equation 4.31 compactly as

$$\mathbf{x} = ||\mathbf{x}|| \, \mathbf{u_x}. \tag{4.33}$$

Here $\mathbf{x}$ is written as the product of its magnitude $||\mathbf{x}||$ and its direction vector $\mathbf{u_x}$. Now we can give an easy procedure to find a vector's direction angles:

(i) find $||\mathbf{x}||$;

(ii) calculate $\mathbf{u_x} = \dfrac{\mathbf{x}}{||\mathbf{x}||}$; and

(iii) take the arc cosines of the elements of $\mathbf{u_x}$.

Step (iii) is often unnecessary; we are usually more interested in the direction vector (unit vector) $\mathbf{u_x}$. Direction vectors are used in materials science in order to study the orientation of crystal lattices and in electromagnetic field theory to characterize the direction of propagation for radar and microwaves. You will find them of inestimable value in your courses on electromagnetic fields and antenna design.

Problem 4.11   Sketch an arbitrary unit vector $\mathbf{u} \in \mathcal{R}^3$. Label the direction cosines and the components of $\mathbf{u}$. ∎

Problem 4.12   Compute the norm and the direction cosines for the vector

$$\mathbf{x} = \begin{bmatrix} 4 \\ 2 \\ 6 \end{bmatrix}. \quad ∎$$

Problem 4.13   Prove that the direction cosines for any vector satisfy the equality

$$\cos^2 \theta_1 + \cos^2 \theta_2 + \cdots + \cos^2 \theta_n = 1.$$

What does this imply about the number of scalars needed to determine a vector $\mathbf{x} \in \mathcal{R}^n$? ∎

Problem 4.14   Astronomers use right ascension, declination, and distance to locate stars. On Figure 4.6 these are, respectively, $-\phi$, $\frac{\pi}{2} - \theta_3$, and $||\mathbf{x}||$. Represent $\mathbf{x} = (x_1, x_2, x_3)$ in terms of ϕ, θ_3, and $||\mathbf{x}||$ only. (That is, find equations that give ϕ, θ_3, and $\|\mathbf{x}\|$ in terms of x_1, x_2, and x_3, and find equations that give x_1, x_2, and x_3 in terms of ϕ, θ_3, and $\|\mathbf{x}\|$.) ∎

Problem 4.15 (MATLAB) Write a MATLAB function that accepts any vector $\mathbf{x} \in \mathcal{R}^n$ and computes its norm and direction cosines. Make $\mathbf{x}$ an input variable, and make $\|\mathbf{x}\|$ and $\mathbf{u_x}$ output variables. ∎

Problem 4.16 Let $\mathbf{x}$ and $\mathbf{y}$ denote two vectors in $\mathcal{R}^n$ with respective direction cosines $(\cos\theta_1, \cos\theta_2, \dots, \cos\theta_n)$ and $(\cos\phi_1, \cos\phi_2, \dots, \cos\phi_n)$. Prove that ψ, the angle between $\mathbf{x}$ and $\mathbf{y}$, is

$$\cos\psi = \cos\theta_1\cos\phi_1 + \cos\theta_2\cos\phi_2 + \cdots + \cos\theta_n\cos\phi_n.$$

Specialize this result to $\mathcal{R}^2$ for insight. ∎

Problem 4.17 Compute the angle between the vectors $\mathbf{x} = \begin{bmatrix} 2 \\ 3 \\ 0 \end{bmatrix}$ and $\begin{bmatrix} 2 \\ 3 \\ 1 \end{bmatrix}$.

Sketch the result. ∎

4.5 Projections

Orthogonality. When the angle between two vectors is $\pi/2$ ($90°$), we say that the vectors are *orthogonal*. A quick look at the definition of angle (Equation 4.28) leads to this equivalent definition for orthogonality:

$$(\mathbf{x}, \mathbf{y}) = 0 \iff \mathbf{x} \text{ and } \mathbf{y} \text{ are orthogonal.} \qquad 4.34$$

For example, in Figure 4.7(a), the vectors $\mathbf{x} = \begin{bmatrix} 3 \\ 1 \end{bmatrix}$ and $\mathbf{y} = \begin{bmatrix} -2 \\ 6 \end{bmatrix}$ are clearly orthogonal, and their inner product is zero:

$$(\mathbf{x}, \mathbf{y}) = 3(-2) + 1(6) = 0.$$

In Figure 4.7(b), the vectors $\mathbf{x} = \begin{bmatrix} 3 \\ 1 \\ 0 \end{bmatrix}$, $\mathbf{y} = \begin{bmatrix} -2 \\ 6 \\ 0 \end{bmatrix}$, and $\mathbf{z} = \begin{bmatrix} 0 \\ 0 \\ 4 \end{bmatrix}$ are

mutually orthogonal, and the inner product between each pair is zero:

$$(\mathbf{x}, \mathbf{y}) = 3(-2) + 1(6) + 0(0) = 0$$

$$(\mathbf{x}, \mathbf{z}) = 3(0) + 1(0) + 0(4) = 0$$

$$(\mathbf{y}, \mathbf{z}) = -2(0) + 6(0) + 0(4) = 0.$$

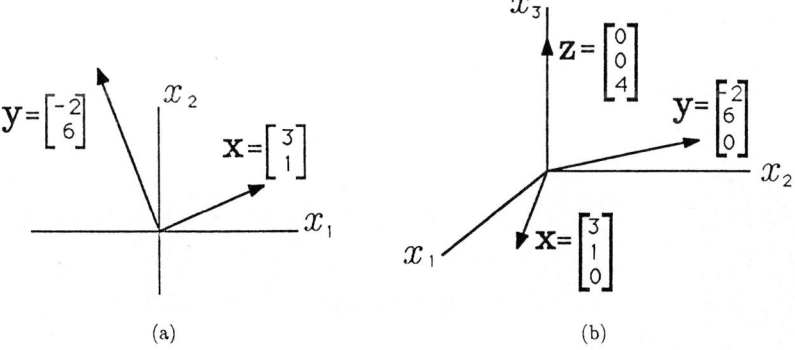

(a)   (b)

Figure 4.7: Orthogonality of Vectors

Problem 4.18   Which of the following pairs of vectors is orthogonal:

(a) $\mathbf{x} = \begin{bmatrix} 1 \\ 0 \\ 1 \end{bmatrix}$, $\mathbf{y} = \begin{bmatrix} 0 \\ 1 \\ 0 \end{bmatrix}$;

(b) $\mathbf{x} = \begin{bmatrix} 1 \\ 1 \\ 0 \end{bmatrix}$, $\mathbf{y} = \begin{bmatrix} 1 \\ 1 \\ 1 \end{bmatrix}$;

(c) $\mathbf{x} = \mathbf{e}_1$, $\mathbf{y} = \mathbf{e}_3$;

(d) $\mathbf{x} = \begin{bmatrix} a \\ b \end{bmatrix}$, $\mathbf{y} = \begin{bmatrix} -b \\ a \end{bmatrix}$? ■

Projection. We can use the inner product to find the projection of one vector onto another as illustrated in Figure 4.8. Geometrically we find the projection of $\mathbf{x}$ onto $\mathbf{y}$ by dropping a perpendicular from the head of $\mathbf{x}$ onto the line containing $\mathbf{y}$. The perpendicular is the dashed line in the figure. The point where the perpendicular intersects $\mathbf{y}$ (or an extension of $\mathbf{y}$) is the projection of $\mathbf{x}$ onto $\mathbf{y}$, or the component of $\mathbf{x}$ along $\mathbf{y}$. Let's call it $\mathbf{z}$.

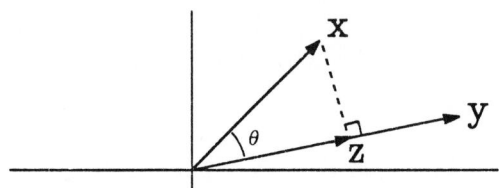

Figure 4.8: Component of One Vector along Another

Problem 4.19 Draw a figure like Figure 4.8 showing the projection of $\mathbf{y}$ onto $\mathbf{x}$. ∎

The vector $\mathbf{z}$ lies along $\mathbf{y}$, so we may write it as the product of its norm $\|\mathbf{z}\|$ and its direction vector $\mathbf{u_y}$:

$$\mathbf{z} = \|\mathbf{z}\|\mathbf{u_y} = \|\mathbf{z}\|\frac{\mathbf{y}}{\|\mathbf{y}\|}. \qquad 4.35$$

But what is norm $\|\mathbf{z}\|$? From Figure 4.8 we see that the vector $\mathbf{x}$ is just $\mathbf{z}$, plus a vector $\mathbf{v}$ that is orthogonal to $\mathbf{y}$:

$$\mathbf{x} = \mathbf{z} + \mathbf{v}, \quad (\mathbf{v}, \mathbf{y}) = 0.$$

Therefore we may write the inner product between $\mathbf{x}$ and $\mathbf{y}$ as

$$(\mathbf{x}, \mathbf{y}) = (\mathbf{z} + \mathbf{v}, \mathbf{y}) = (\mathbf{z}, \mathbf{y}) + (\mathbf{v}, \mathbf{y}) = (\mathbf{z}, \mathbf{y}). \qquad 4.36$$

But because $\mathbf{z}$ and $\mathbf{y}$ both lie along $\mathbf{y}$, we may write the inner product $(\mathbf{x}, \mathbf{y})$ as

$$(\mathbf{x}, \mathbf{y}) = (\mathbf{z}, \mathbf{y}) = (\|\mathbf{z}\|\mathbf{u_y}, \|\mathbf{y}\|\mathbf{u_y}) = \|\mathbf{z}\| \, \|\mathbf{y}\|(\mathbf{u_y}, \mathbf{u_y})$$
$$= \|\mathbf{z}\| \, \|\mathbf{y}\| \, \|\mathbf{u_y}\|^2 = \|\mathbf{z}\| \, \|\mathbf{y}\|. \qquad 4.37$$

From this equation we may solve for $||\mathbf{z}|| = \frac{(\mathbf{x},\mathbf{y})}{||\mathbf{y}||}$ and substitute $||\mathbf{z}||$ into Equation 4.35 to write $\mathbf{z}$ as

$$
\begin{aligned}
\mathbf{z} &= ||\mathbf{z}|| \frac{\mathbf{y}}{||\mathbf{y}||} \\
&= \frac{(\mathbf{x},\mathbf{y})}{||\mathbf{y}||} \frac{\mathbf{y}}{||\mathbf{y}||} = \frac{(\mathbf{x},\mathbf{y})}{(\mathbf{y},\mathbf{y})} \mathbf{y}.
\end{aligned}
\qquad 4.38
$$

Equation 4.38 is what we wanted—an expression for the projection of $\mathbf{x}$ onto $\mathbf{y}$ in terms of $\mathbf{x}$ and $\mathbf{y}$.

Problem 4.20 Show that $||\mathbf{z}||$ and $\mathbf{z}$ may be written in terms of $\cos\theta$ for θ as illustrated in Figure 4.8:

$$||\mathbf{z}|| = ||\mathbf{x}|| \cos\theta$$

$$\mathbf{z} = \frac{||\mathbf{x}|| \cos\theta}{||\mathbf{y}||} \mathbf{y}. \quad \blacksquare$$

Orthogonal Decomposition. You already know how to decompose a vector in terms of the unit coordinate vectors,

$$\mathbf{x} = (\mathbf{x},\mathbf{e}_1)\mathbf{e}_1 + (\mathbf{x},\mathbf{e}_2)\mathbf{e}_2 + \cdots + (\mathbf{x},\mathbf{e}_n)\mathbf{e}_n.$$

In this equation, $(\mathbf{x},\mathbf{e}_k)\mathbf{e}_k$ is the component of $\mathbf{x}$ along $\mathbf{e}_k$, or the projection of $\mathbf{x}$ onto $\mathbf{e}_k$, but the set of unit coordinate vectors is not the only possible basis for decomposing a vector. Let's consider an arbitrary pair of orthogonal vectors $\mathbf{x}$ and $\mathbf{y}$:

$$(\mathbf{x},\mathbf{y}) = 0.$$

The sum of $\mathbf{x}$ and $\mathbf{y}$ produces a new vector $\mathbf{w}$, illustrated in Figure 4.9, where we have used a two-dimensional drawing to represent n dimensions. The norm squared of $\mathbf{w}$ is

$$
\begin{aligned}
||\mathbf{w}||^2 &= (\mathbf{w},\mathbf{w}) = [(\mathbf{x}+\mathbf{y}),(\mathbf{x}+\mathbf{y})] = (\mathbf{x},\mathbf{x}) + (\mathbf{x},\mathbf{y}) + (\mathbf{y},\mathbf{x}) + (\mathbf{y},\mathbf{y}) \\
&= ||\mathbf{x}||^2 + ||\mathbf{y}||^2.
\end{aligned}
\qquad 4.39
$$

This is the Pythagorean theorem in n dimensions! The length squared of $\mathbf{w}$ is just the sum of the squares of the lengths of its two orthogonal components.

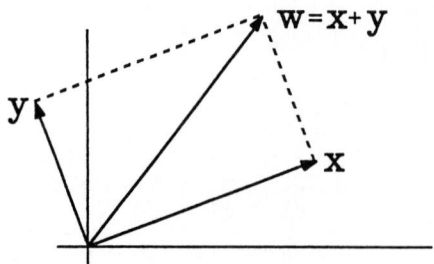

Figure 4.9: Sum of Orthogonal Vectors

The projection of $\mathbf{w}$ onto $\mathbf{x}$ is $\mathbf{x}$, and the projection of $\mathbf{w}$ onto $\mathbf{y}$ is $\mathbf{y}$:

$$\mathbf{w} = (1)\mathbf{x} + (1)\mathbf{y}. \qquad 4.40$$

If we scale $\mathbf{w}$ by a to produce the vector $\mathbf{z} = a\mathbf{w}$, the orthogonal decomposition of $\mathbf{z}$ is

$$\mathbf{z} = a\mathbf{w} = (a)\mathbf{x} + (a)\mathbf{y}. \qquad 4.41$$

Let's turn this argument around. Instead of building $\mathbf{w}$ from orthogonal vectors $\mathbf{x}$ and $\mathbf{y}$, let's begin with arbitrary $\mathbf{w}$ and $\mathbf{x}$ and see whether we can compute an orthogonal decomposition. The projection of $\mathbf{w}$ onto $\mathbf{x}$ is found from Equation 4.38:

$$\mathbf{w_x} = \frac{(\mathbf{w}, \mathbf{x})}{(\mathbf{x}, \mathbf{x})} \mathbf{x}. \qquad 4.42$$

But there must be another component of $\mathbf{w}$ such that $\mathbf{w}$ is equal to the sum of the components. Let's call the unknown component $\mathbf{w_y}$. Then

$$\mathbf{w} = \mathbf{w_x} + \mathbf{w_y}. \qquad 4.43$$

Now, since we know $\mathbf{w}$ and $\mathbf{w_x}$ already, we find $\mathbf{w_y}$ to be

$$\mathbf{w_y} = \mathbf{w} - \mathbf{w_x} = \mathbf{w} - \frac{(\mathbf{w}, \mathbf{x})}{(\mathbf{x}, \mathbf{x})} \mathbf{x}. \qquad 4.44$$

Interestingly, the way we have decomposed $\mathbf{w}$ will always produce $\mathbf{w_x}$ and $\mathbf{w_y}$ orthogonal to each other. Let's check this:

$$
\begin{aligned}
(\mathbf{w_x}, \mathbf{w_y}) &= \left(\frac{(\mathbf{w}, \mathbf{x})}{(\mathbf{x}, \mathbf{x})} \, \mathbf{x}, \ \mathbf{w} - \frac{(\mathbf{w}, \mathbf{x})}{(\mathbf{x}, \mathbf{x})} \, \mathbf{x} \right) \\
&= \frac{(\mathbf{w}, \mathbf{x})}{(\mathbf{x}, \mathbf{x})} \, (\mathbf{x}, \mathbf{w}) - \frac{(\mathbf{w}, \mathbf{x})^2}{(\mathbf{x}, \mathbf{x})^2} \, (\mathbf{x}, \mathbf{x}) \\
&= \frac{(\mathbf{w}, \mathbf{x})^2}{(\mathbf{x}, \mathbf{x})} - \frac{(\mathbf{w}, \mathbf{x})^2}{(\mathbf{x}, \mathbf{x})} \\
&= 0.
\end{aligned}
$$

4.45

To summarize, we have taken two arbitrary vectors, $\mathbf{w}$ and $\mathbf{x}$, and decomposed $\mathbf{w}$ into a component in the direction of $\mathbf{x}$ and a component orthogonal to $\mathbf{x}$.

4.6 Other Norms

Sometimes we find it useful to use a different definition of distance, corresponding to an alternate norm for vectors. For example, consider the *1-norm* defined as

$$ \|\, \mathbf{x} \,\|_1 = \left(|x_1| + |x_2| + \cdots + |x_n| \right), \qquad 4.46 $$

where $|x_i|$ is the magnitude of component x_i. There is also the *sup-norm*, the "supremum" or maximum of the components $x_1, \ldots, x_n$:

$$ \|\mathbf{x}\|_{\mathrm{sup}} = \max\left(|x_1|, |x_2|, \ldots, |x_n| \right). \qquad 4.47 $$

The following examples illustrate what the Euclidean norm, the 1-norm, and the sup-norm look like for typical vectors.

Example 4.4. Consider the vector $\mathbf{x} = \begin{bmatrix} -3 \\ 1 \\ 2 \end{bmatrix}$. Then

(i) $\|\mathbf{x}\| = \left[(-3)^2 + (1)^2 + (2)^2 \right]^{1/2} = (14)^{1/2}$;

(ii) $\|\mathbf{x}\|_1 = \left(|-3| + |1| + |2| \right) = 6$; and

(iii) $\|\mathbf{x}\|_{\mathrm{sup}} = \max\left(|-3|, |1|, |2| \right) = 3.$ ☐

Example 4.5. Figure 4.10 shows the locus of two-component vectors $\mathbf{x} = \begin{bmatrix} x_1 \\ x_2 \end{bmatrix}$ with the property that $\|\mathbf{x}\| = 1$, $\|\mathbf{x}\|_1 = 1$, or $\|\mathbf{x}\|_{\text{sup}} = 1$. $\square$

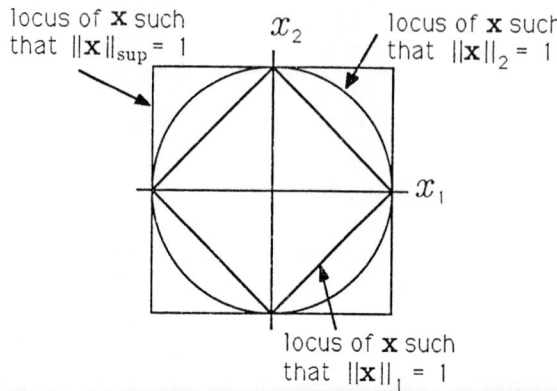

Figure 4.10: Locus of Two-Dimensional Vectors Whose Various Norms Are 1

The next example shows how the 1-norm is an important part of city life.

Example 4.6. The city of Metroville was laid out by mathematicians as shown in Figure 4.11. A person at the intersection of Avenue 0 and Street -2 (point A) is clearly two blocks from the center of town (point C). This is consistent with both the Euclidean norm

$$\| A \| = \sqrt{0^2 + (-2)^2} = \sqrt{4} = 2$$

and the 1-norm

$$\| A \|_1 = \big(|0| + |-2| \big) = 2.$$

But how far from the center of town is point B at the intersection of Avenue -2 and Street 1? According to the Euclidean norm, the distance is

$$\| B \| = \sqrt{(-2)^2 + (1)^2} = \sqrt{5}.$$

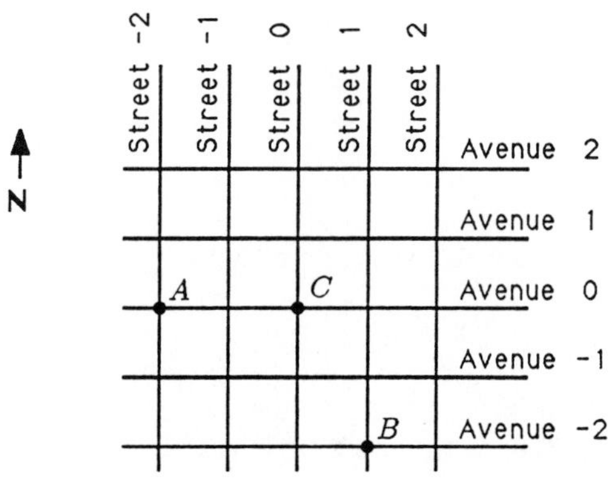

Figure 4.11: Metroville, U.S.A.

While it is true that point B is $\sqrt{5}$ blocks from C, it is also clear that the trip would be three blocks by any of the three shortest routes on roads. The appropriate norm is the 1-norm:

$$\| B \|_1 = (|-2| + |1|) = 3. \quad \square$$

Even more generally, we can define a norm for each value of p from 1 to infinity. The so-called *p-norm* is

$$\| \mathbf{x} \|_p = \left(|x_1|^p + |x_2|^p + \cdots + |x_n|^p \right)^{1/p}. \qquad 4.48$$

Problem 4.21 Show that the Euclidean norm is the same as the p-norm with $p = 2$ and that the 1-norm is the p-norm with $p = 1$. (It can also be shown that the sup-norm is like a p-norm with $p = \infty$.) ∎

DEMO 4.1 (MATLAB). From the command level of MATLAB, type the following lines:

```
≫ x = [1;3;-2;4]
≫ y = [0;1;2;-0.5]
≫ x - y
```

Check to see whether the answer agrees with the definition of vector subtraction. Now type

```
≫ a = -1.5
≫ a * x
```

Check the answer to see whether it agrees with the definition of scalar multiplication. Now type

```
≫ x' * y
```

This is how MATLAB does the inner product. Check the result. Type

```
≫ norm(y)
≫ sqrt(y' * y)
```

These two ways of computing the Euclidean norm should agree. Check the 1-norm and the sup-norm:

```
≫ norm(y,1)
≫ norm(y,inf)
```

Now type your own MATLAB expression to find the cosine of the angle between vectors x and y. Put the result in variable t. Then find the angle θ by typing

```
≫ theta = acos(t)
```

The angle θ is in radians. You may convert it to degrees if you wish by multiplying it by $180/\pi$:

```
≫ theta = theta * (180/pi)   □
```

4.7 Matrices

The word *matrix* dates at least to the thirteenth century, when it was used to describe the rectangular copper tray, or matrix, that held individual leaden letters that were packed into the matrix to form a page of composed text. Each letter in the matrix, call it a_{ij}, occupied a unique position (i, j) in the matrix. In modern day mathematical parlance, a matrix is a collection of numbers arranged in a two-dimensional array (a rectangle). We indicate a matrix with a boldfaced capital letter and the constituent elements with double subscripts for the row and column:

$$\mathbf{A} = \begin{bmatrix} a_{11} & a_{12} & a_{13} & \cdots & a_{1n} \\ a_{21} & a_{22} & a_{23} & \cdots & a_{2n} \\ a_{31} & a_{32} & a_{33} & \cdots & a_{3n} \\ \vdots & \vdots & \vdots & & \vdots \\ a_{m1} & a_{m2} & a_{m3} & \cdots & a_{mn} \end{bmatrix} \quad \leftarrow \text{second row}$$

$\uparrow$
third column

4.49

In this equation $\mathbf{A}$ is an $m \times n$ matrix, meaning that $\mathbf{A}$ has m horizontal rows and n vertical columns. As an extension of the previously used notation, we write $\mathbf{A} \in \mathcal{R}^{m \times n}$ to show that $\mathbf{A}$ is a matrix of size $m \times n$ with $a_{ij} \in \mathcal{R}$. The scalar element a_{ij} is located in the matrix at the ith row and the jth column. For example, a_{23} is located in the second row and the third column as illustrated in Equation 4.49.

The *main diagonal* of any matrix consists of the elements a_{ii}. (The two subscripts are equal.) The main diagonal runs southeast from the top left corner of the matrix, but it does not end in the lower right corner unless the matrix is square ($\in \mathcal{R}^{m \times m}$).

The *transpose* of a matrix $\mathbf{A} \in \mathcal{R}^{m \times n}$ is another matrix $\mathbf{B}$ whose element in row j and column i is $b_{ji} = a_{ij}$ for $1 \leq i \leq m$ and $1 \leq j \leq n$. We write $\mathbf{B} = \mathbf{A}^T$ to indicate that $\mathbf{B}$ is the transpose of $\mathbf{A}$. In MATLAB, transpose is denoted by A'. A more intuitive way of describing the transpose operation is to say that it flips the matrix about its main diagonal so that rows become columns and columns become rows.

Problem 4.22 If $\mathbf{A} \in \mathcal{R}^{m \times n}$, then $\mathbf{A}^T \in \underline{\ ?\ }$. Find the transpose of the matrix

$$\mathbf{A} = \begin{bmatrix} 2 & 1 \\ 5 & 4 \\ 7 & 9 \end{bmatrix}. \quad \blacksquare$$

Matrix Addition and Scalar Multiplication. Two matrices of the same size (in both dimensions) may be added or subtracted in the same way as vectors, by adding or subtracting the corresponding elements. The equation $\mathbf{C} = \mathbf{A} \pm \mathbf{B}$ means that for each i and j, $c_{ij} = a_{ij} \pm b_{ij}$. Scalar multiplication of a matrix multiplies each element of the matrix by the scalar:

$$a\mathbf{X} = \begin{bmatrix} ax_{11} & ax_{12} & ax_{13} & \cdots & ax_{1n} \\ ax_{21} & ax_{22} & ax_{23} & \cdots & ax_{2n} \\ ax_{31} & ax_{32} & ax_{33} & \cdots & ax_{3n} \\ \vdots & \vdots & \vdots & & \vdots \\ ax_{m1} & ax_{m2} & ax_{m3} & \cdots & ax_{mn} \end{bmatrix}. \qquad 4.50$$

Matrix Multiplication. A vector can be considered a matrix with only one column. Thus we intentionally blur the distinction between $\mathcal{R}^{n\times 1}$ and $\mathcal{R}^n$. Also a matrix can be viewed as a collection of vectors, each column of the matrix being a vector:

$$\mathbf{A} = \begin{bmatrix} | & | & & | \\ \mathbf{a}_1 & \mathbf{a}_2 & \cdots & \mathbf{a}_n \\ | & | & & | \end{bmatrix} = \begin{bmatrix} \begin{bmatrix} a_{11} \\ a_{21} \\ \vdots \\ a_{m1} \end{bmatrix} & \begin{bmatrix} a_{12} \\ a_{22} \\ \vdots \\ a_{m2} \end{bmatrix} & \cdots & \begin{bmatrix} a_{1n} \\ a_{2n} \\ \vdots \\ a_{mn} \end{bmatrix} \end{bmatrix}. \qquad 4.51$$

In the transpose operation, columns become rows and vice versa. The transpose of an $n \times 1$ matrix, a *column vector*, is a $1 \times n$ matrix, a *row vector*:

$$\mathbf{x} = \begin{bmatrix} x_1 \\ x_2 \\ \vdots \\ x_n \end{bmatrix}; \qquad \mathbf{x}^T = \begin{bmatrix} x_1 & x_2 & \cdots & x_n \end{bmatrix}. \qquad 4.52$$

Now we can define matrix-matrix multiplication in terms of inner products of vectors. Let's begin with matrices $\mathbf{A} \in \mathcal{R}^{m\times n}$ and $\mathbf{B} \in \mathcal{R}^{n\times p}$. To find the product $\mathbf{AB}$, first divide each matrix into column vectors and row vectors as follows:

$$
\mathbf{A} = \begin{bmatrix} \begin{bmatrix} | \\ \mathbf{a}_1 \\ | \end{bmatrix} & \begin{bmatrix} | \\ \mathbf{a}_2 \\ | \end{bmatrix} & \cdots & \begin{bmatrix} | \\ \mathbf{a}_n \\ | \end{bmatrix} \end{bmatrix} = \begin{bmatrix} [- & \boldsymbol{\alpha}_1^T & -] \\ [- & \boldsymbol{\alpha}_2^T & -] \\ & \vdots & \\ [- & \boldsymbol{\alpha}_m^T & -] \end{bmatrix} \qquad 4.53
$$

$$
\mathbf{B} = \begin{bmatrix} \begin{bmatrix} | \\ \mathbf{b}_1 \\ | \end{bmatrix} & \begin{bmatrix} | \\ \mathbf{b}_2 \\ | \end{bmatrix} & \cdots & \begin{bmatrix} | \\ \mathbf{b}_p \\ | \end{bmatrix} \end{bmatrix} = \begin{bmatrix} [- & \boldsymbol{\beta}_1^T & -] \\ [- & \boldsymbol{\beta}_2^T & -] \\ & \vdots & \\ [- & \boldsymbol{\beta}_n^T & -] \end{bmatrix}.
$$

Thus $\mathbf{a}_i$ is the ith column of $\mathbf{A}$ and $\boldsymbol{\alpha}_j^T$ is the jth row of $\mathbf{A}$. For matrix multiplication to be defined, the width of the first matrix must match the length of the second one so that all rows $\boldsymbol{\alpha}_i^T$ and columns $\mathbf{b}_i$ have the same number of elements n. The matrix product, $\mathbf{C} = \mathbf{AB}$, is an $m \times p$ matrix defined by its elements as $c_{ij} = (\boldsymbol{\alpha}_i, \mathbf{b}_j)$. In words, each element of the product matrix, c_{ij}, is the inner product of the ith row of the first matrix and the jth column of the second matrix.

For n-vectors $\mathbf{x}$ and $\mathbf{y}$, the matrix product $\mathbf{x}^T\mathbf{y}$ takes on a special significance. The product is, of course, a 1×1 matrix (a scalar). The special significance is that $\mathbf{x}^T\mathbf{y}$ is the inner product of $\mathbf{x}$ and $\mathbf{y}$:

$$
\begin{bmatrix} [- & \mathbf{x}^T & -] \end{bmatrix} \begin{bmatrix} | \\ \mathbf{y} \\ | \end{bmatrix} = (\mathbf{x}, \mathbf{y}). \qquad 4.54
$$

Thus the notation $\mathbf{x}^T\mathbf{y}$ is often used in place of $(\mathbf{x}, \mathbf{y})$. Recall from Demo 4.1 that MATLAB uses x'*y to denote inner product.

Another special case of matrix multiplication is the *outer product*. Like the inner product, it involves two vectors, but this time the result is a matrix:

$$\mathbf{A} = \begin{bmatrix} | \\ \mathbf{x} \\ | \end{bmatrix} \begin{bmatrix} - & \mathbf{y}^T & - \end{bmatrix} = \begin{bmatrix} x_1 y_1 & x_1 y_2 & \cdots & x_1 y_n \\ x_2 y_1 & x_2 y_2 & \cdots & x_2 y_n \\ \vdots & \vdots & & \vdots \\ x_m y_1 & x_m y_2 & \cdots & x_m y_n \end{bmatrix}.$$

In the outer product, the inner products that define its elements are between one-dimensional row vectors of $\mathbf{x}$ and one-dimensional column vectors of $\mathbf{y}^T$, meaning the (i, j) element of $\mathbf{A}$ is $x_i y_j$.

Problem 4.23 Find $\mathbf{C} = \mathbf{AB}$ where $\mathbf{A}$ and $\mathbf{B}$ are given by

(a) $\mathbf{A} = \begin{bmatrix} 1 & -1 & 2 \\ 3 & 0 & 5 \end{bmatrix}$, $\quad \mathbf{B} = \begin{bmatrix} 0 & -2 & 1 & -3 \\ 4 & 2 & 2 & 0 \\ 2 & -2 & 3 & 1 \end{bmatrix}$;

(b) $\mathbf{A} = \begin{bmatrix} 1 & 0 \\ 0 & 1 \end{bmatrix}$, $\quad \mathbf{B} = \begin{bmatrix} 1 & 2 & 3 & 4 \\ 5 & 6 & 7 & 8 \end{bmatrix}$;

(c) $\mathbf{A} = \begin{bmatrix} 1 & -1 & -1 \\ 1 & -1 & 1 \\ 1 & 1 & 1 \end{bmatrix}$, $\quad \mathbf{B} = \begin{bmatrix} 0 & 3 & 6 \\ 1 & 4 & 7 \\ 2 & 5 & 8 \end{bmatrix}$. ∎

There are several other equivalent ways to define matrix multiplication, and a careful study of the following discussion should improve your understanding of matrix multiplication. Consider $\mathbf{A} \in \mathcal{R}^{m \times n}$, $\mathbf{B} \in \mathcal{R}^{n \times p}$, and $\mathbf{C} = \mathbf{AB}$ so that $\mathbf{C} \in \mathcal{R}^{m \times p}$. In pictures, we have

$$m \begin{bmatrix} & & p & \\ & & \mathbf{C} & \\ & & & \end{bmatrix} = m \begin{bmatrix} & n & \\ & \mathbf{A} & \\ & & \end{bmatrix} \begin{bmatrix} p \\ \mathbf{B} \end{bmatrix} n. \qquad 4.55$$

In our definition, we represent $\mathbf{C}$ on an entry-by-entry basis as

$$c_{ij} = (\boldsymbol{\alpha}_i, \mathbf{b}_j) = \sum_{k=1}^{n} a_{ik} b_{kj}. \qquad 4.56$$

In pictures,

$$
\begin{bmatrix} & & \\ & c_{ij} & \\ & & \end{bmatrix} = \begin{bmatrix} & & \\ [- & \mathbf{a}_i^{\ T} & -] \\ & & \end{bmatrix} \begin{bmatrix} & | & \\ & \mathbf{b}_j & \\ & | & \end{bmatrix} . \qquad 4.57
$$

You will prove in Problem 4.24 that we can also represent $\mathbf{C}$ on a column basis:

$$
\mathbf{c}_j = \sum_{k=1}^{n} \mathbf{a}_k b_{kj} . \qquad 4.58
$$

In pictures,

$$
\begin{bmatrix} & \begin{bmatrix} | \\ \mathbf{c}_j \\ | \end{bmatrix} & \end{bmatrix} = \begin{bmatrix} | \\ \mathbf{a}_1 \\ | \end{bmatrix} \begin{bmatrix} | \\ \mathbf{a}_2 \\ | \end{bmatrix} \cdots \begin{bmatrix} | \\ \mathbf{a}_n \\ | \end{bmatrix} \begin{bmatrix} | \\ \mathbf{b}_j \\ | \end{bmatrix} .
$$

$$4.59$$

Finally, $\mathbf{C}$ can be represented as a sum of matrices, each matrix being an outer product:

$$
\mathbf{C} = \sum_{i=1}^{n} \mathbf{a}_i \boldsymbol{\beta}_i^T \qquad 4.60
$$

$$
\begin{bmatrix} & & \\ & \mathbf{C} & \\ & & \end{bmatrix} = \begin{bmatrix} | \\ \mathbf{a}_1 \\ | \end{bmatrix} [- \ \boldsymbol{\beta}_1^T \ -] + \begin{bmatrix} | \\ \mathbf{a}_2 \\ | \end{bmatrix} [- \ \boldsymbol{\beta}_2^T \ -] + \cdots .
$$

$$4.61$$

A numerical example should help clarify these three methods.

Example 4.7. Let

$$A = \begin{bmatrix} 1 & 2 & 1 & 3 \\ 2 & 1 & 2 & 4 \\ 3 & 3 & 2 & 1 \end{bmatrix}, \qquad B = \begin{bmatrix} 1 & 2 & 1 \\ 2 & 2 & 1 \\ 1 & 3 & 2 \\ 2 & 1 & 1 \end{bmatrix}.$$

Using the first method of matrix multiplication, on an entry-by-entry basis, we have

$$c_{ij} = \sum_{k=1}^{4} a_{ik} b_{kj}$$

or

$$\mathbf{C} = \begin{bmatrix} (1 \cdot 1 + 2 \cdot 2 + 1 \cdot 1 + 3 \cdot 2) & (1 \cdot 2 + 2 \cdot 2 + 1 \cdot 3 + 3 \cdot 1) \\ (2 \cdot 1 + 1 \cdot 2 + 2 \cdot 1 + 4 \cdot 2) & (2 \cdot 2 + 1 \cdot 2 + 2 \cdot 3 + 4 \cdot 1) \\ (3 \cdot 1 + 3 \cdot 2 + 2 \cdot 1 + 1 \cdot 2) & (3 \cdot 2 + 3 \cdot 2 + 2 \cdot 3 + 1 \cdot 1) \end{bmatrix}$$

$$\begin{bmatrix} (1 \cdot 1 + 2 \cdot 1 + 1 \cdot 2 + 3 \cdot 1) \\ (2 \cdot 1 + 1 \cdot 1 + 2 \cdot 2 + 4 \cdot 1) \\ (3 \cdot 1 + 3 \cdot 1 + 2 \cdot 2 + 1 \cdot 1) \end{bmatrix}$$

or

$$\mathbf{C} = \begin{bmatrix} 12 & 12 & 8 \\ 14 & 16 & 11 \\ 13 & 19 & 11 \end{bmatrix}.$$

On a column basis,

$$\mathbf{c}_j = \sum_{k=1}^{4} \mathbf{a}_k b_{kj}$$

$$\mathbf{c}_1 = \begin{bmatrix} 1 \\ 2 \\ 3 \end{bmatrix} 1 + \begin{bmatrix} 2 \\ 1 \\ 3 \end{bmatrix} 2 + \begin{bmatrix} 1 \\ 2 \\ 2 \end{bmatrix} 1 + \begin{bmatrix} 3 \\ 4 \\ 1 \end{bmatrix} 2; \quad \mathbf{c}_2 = \begin{bmatrix} 1 \\ 2 \\ 3 \end{bmatrix} 2 + \begin{bmatrix} 2 \\ 1 \\ 3 \end{bmatrix} 2 + \begin{bmatrix} 1 \\ 2 \\ 2 \end{bmatrix} 3 + \begin{bmatrix} 3 \\ 4 \\ 1 \end{bmatrix} 1;$$

$$\mathbf{c}_3 = \begin{bmatrix} 1 \\ 2 \\ 3 \end{bmatrix} 1 + \begin{bmatrix} 2 \\ 1 \\ 3 \end{bmatrix} 1 + \begin{bmatrix} 1 \\ 2 \\ 2 \end{bmatrix} 2 + \begin{bmatrix} 3 \\ 4 \\ 1 \end{bmatrix} 1.$$

Collecting terms together, we have

$$\mathbf{C} = [\mathbf{c}_1 \quad \mathbf{c}_2 \quad \mathbf{c}_3]$$

$$= \begin{bmatrix} (1 \cdot 1 + 2 \cdot 2 + 1 \cdot 1 + 3 \cdot 2) & (1 \cdot 2 + 2 \cdot 2 + 1 \cdot 3 + 3 \cdot 1) \\ (2 \cdot 1 + 1 \cdot 2 + 2 \cdot 1 + 4 \cdot 2) & (2 \cdot 2 + 1 \cdot 2 + 2 \cdot 3 + 4 \cdot 1) \\ (3 \cdot 1 + 3 \cdot 2 + 2 \cdot 1 + 1 \cdot 2) & (3 \cdot 2 + 3 \cdot 2 + 2 \cdot 3 + 1 \cdot 1) \end{bmatrix}$$
$$\begin{matrix} (1 \cdot 1 + 2 \cdot 1 + 1 \cdot 2 + 3 \cdot 1) \\ (2 \cdot 1 + 1 \cdot 1 + 2 \cdot 2 + 4 \cdot 1) \\ (3 \cdot 1 + 3 \cdot 1 + 2 \cdot 2 + 1 \cdot 1) \end{matrix} \quad .$$

On a matrix-by-matrix basis,

$$\mathbf{C} = \sum_{i=1}^{4} \mathbf{a}_i \boldsymbol{\beta}_i^T$$

$$\mathbf{C} = \begin{bmatrix} 1 \\ 2 \\ 3 \end{bmatrix} [1\ 2\ 1] + \begin{bmatrix} 2 \\ 1 \\ 3 \end{bmatrix} [2\ 2\ 1] + \begin{bmatrix} 1 \\ 2 \\ 2 \end{bmatrix} [1\ 3\ 2] + \begin{bmatrix} 3 \\ 4 \\ 1 \end{bmatrix} [2\ 1\ 1]$$

$$= \begin{bmatrix} 1 \cdot 1 & 1 \cdot 2 & 1 \cdot 1 \\ 2 \cdot 1 & 2 \cdot 2 & 2 \cdot 1 \\ 3 \cdot 1 & 3 \cdot 2 & 3 \cdot 1 \end{bmatrix} + \begin{bmatrix} 2 \cdot 2 & 2 \cdot 2 & 2 \cdot 1 \\ 1 \cdot 2 & 1 \cdot 2 & 1 \cdot 1 \\ 3 \cdot 2 & 3 \cdot 2 & 3 \cdot 1 \end{bmatrix} + \begin{bmatrix} 1 \cdot 1 & 1 \cdot 3 & 1 \cdot 2 \\ 2 \cdot 1 & 2 \cdot 3 & 2 \cdot 2 \\ 2 \cdot 1 & 2 \cdot 3 & 2 \cdot 2 \end{bmatrix}$$
$$+ \begin{bmatrix} 3 \cdot 2 & 3 \cdot 1 & 3 \cdot 1 \\ 4 \cdot 2 & 4 \cdot 1 & 4 \cdot 1 \\ 1 \cdot 2 & 1 \cdot 1 & 1 \cdot 1 \end{bmatrix}$$

$$= \begin{bmatrix} (1 \cdot 1 + 2 \cdot 2 + 1 \cdot 1 + 3 \cdot 2) & (1 \cdot 2 + 2 \cdot 2 + 1 \cdot 3 + 3 \cdot 1) \\ (2 \cdot 1 + 1 \cdot 2 + 2 \cdot 1 + 4 \cdot 2) & (2 \cdot 2 + 1 \cdot 2 + 2 \cdot 3 + 4 \cdot 1) \\ (3 \cdot 1 + 3 \cdot 2 + 2 \cdot 1 + 1 \cdot 2) & (3 \cdot 2 + 3 \cdot 2 + 2 \cdot 3 + 1 \cdot 1) \end{bmatrix}$$
$$\begin{matrix} (1 \cdot 1 + 2 \cdot 1 + 1 \cdot 2 + 3 \cdot 1) \\ (2 \cdot 1 + 1 \cdot 1 + 2 \cdot 2 + 4 \cdot 1) \\ (3 \cdot 1 + 3 \cdot 1 + 2 \cdot 2 + 1 \cdot 1) \end{matrix} \quad ,$$

as we had in each of the other cases. Thus we see that the methods are equivalent—simply different ways of organizing the same computation! □

Problem 4.24 Prove that Equations 4.56, 4.58, and 4.60 are equivalent definitions of matrix multiplication. That is, if $C = AB$ where $A \in \mathcal{R}^{m \times n}$ and $B \in \mathcal{R}^{n \times p}$, show that the matrix-matrix product can also be defined by

$$c_{ij} = \sum_{k=1}^{n} a_{ik} b_{kj},$$

and, if c_k is the kth column of C and a_k is the kth column of A, then

$$c_j = \sum_{k=1}^{n} a_k b_{kj}.$$

Show that the matrix C may also be written as the "sum of outer products"

$$C = \sum_{k=1}^{n} a_k \beta_k^T.$$

Write out the elements in a typical outer product $a_k \beta_k^T$. ∎

Problem 4.25 Given $A \in \mathcal{R}^{m \times n}$, $B \in \mathcal{R}^{p \times q}$, and $C \in \mathcal{R}^{r \times s}$, for each of the following postulates, either prove that it is true or give a counterexample showing that it is false:

(a) $(A^T)^T = A$.

(b) $AB = BA$ when $n = p$ and $m = q$. Is matrix multiplication commutative?

(c) $A(B + C) = AB + AC$ when $n = p = r$ and $q = s$. Is matrix multiplication distributive over addition?

(d) $(AB)^T = B^T A^T$ when $n = p$.

(e) $(AB)C = A(BC)$ when $n = p$ and $q = r$. Is matrix multiplication associative? ∎

Example 4.8 (Rotation). We know from Chapter 1 that a complex number $z_1 = x_1 + jy_1$ may be rotated by angle θ in the complex plane by forming the product

$$z_2 = e^{j\theta} z_1.$$

When written out, the real and imaginary parts of z_2 are

$$z_2 = (\cos\theta + j\sin\theta)(x_1 + jy_1)$$
$$= (\cos\theta)x_1 - (\sin\theta)y_1 + j[(\sin\theta)x_1 + (\cos\theta)y_1].$$

If the real and imaginary parts of z_1 and z_2 are organized into vectors $\mathbf{z}_1$ and $\mathbf{z}_2$ as in Chapter 1, then rotation may be carried out with the matrix-vector multiply

$$\mathbf{z}_2 = \begin{bmatrix} x_2 \\ y_2 \end{bmatrix} = \begin{bmatrix} \cos\theta & -\sin\theta \\ \sin\theta & \cos\theta \end{bmatrix} \begin{bmatrix} x_1 \\ y_1 \end{bmatrix}.$$

We call the matrix $\mathbf{R}(\theta) = \begin{bmatrix} \cos\theta & -\sin\theta \\ \sin\theta & \cos\theta \end{bmatrix}$ a *rotation matrix*. $\square$

Problem 4.26 Let $\mathbf{R}(\theta)$ denote a 2×2 rotation matrix. Prove and interpret the following two properties:

(a) $\mathbf{R}^T(\theta) = \mathbf{R}(-\theta)$;

(b) $\mathbf{R}^T(\theta)\mathbf{R}(\theta) = \mathbf{R}(\theta)\mathbf{R}^T(\theta) = \begin{bmatrix} 1 & 0 \\ 0 & 1 \end{bmatrix}$. ∎

4.8 Solving Linear Systems of Equations

We are now equipped to set up systems of linear equations as matrix-vector equations so that they can be solved in a standard way on a computer. Suppose we want to solve the equations from Example 4.1 for x_1 and x_2 using a computer. Recall that Equations 4.1 and 4.2 are

$$x_1 + x_2 = 85 \tag{4.62}$$

$$\frac{x_1}{1.2} = \frac{x_2}{1.5 - 1.2}.$$

The first step is to arrange each equation with all references to x_1 in the first column, all references to x_2 in the second column, etc., and all constants on the right-hand side:

$$x_1 + x_2 = 85 \tag{4.63}$$

$$0.3x_1 - 1.2x_2 = 0.$$

Then the equations can be converted to a single matrix-vector equation. The coefficients form a matrix, keeping their same relative positions, and the variables and constants each form a vector:

$$\begin{bmatrix} 1 & 1 \\ 0.3 & -1.2 \end{bmatrix} \begin{bmatrix} x_1 \\ x_2 \end{bmatrix} = \begin{bmatrix} 85 \\ 0 \end{bmatrix}. \qquad 4.64$$

Problem 4.27 Verify by the rules of matrix multiplication that the system of equations in 4.63 is equivalent to the matrix equation in 4.64. ∎

Equation 4.64 is of the general form

$$\mathbf{A}\mathbf{x} = \mathbf{b} \qquad 4.65$$

where in this case

$$\mathbf{A} = \begin{bmatrix} 1 & 1 \\ 0.3 & -1.2 \end{bmatrix}, \qquad \mathbf{x} = \begin{bmatrix} x_1 \\ x_2 \end{bmatrix}, \qquad \mathbf{b} = \begin{bmatrix} 85 \\ 0 \end{bmatrix}. \qquad 4.66$$

Given any $\mathbf{A} \in \mathcal{R}^{n \times n}$ and $\mathbf{b} \in \mathcal{R}^n$, MATLAB can solve Equation 4.65 for $\mathbf{x}$ (as long as a solution exists). Key ideas in the solution process are the *identity* matrix and the *inverse* of a matrix.

When the matrix $\mathbf{A}$ is the 1×1 matrix a, the vector $\mathbf{x}$ is the 1-vector x, and the vector $\mathbf{b}$ is the 1-vector b, then the matrix equation $\mathbf{A}\mathbf{x} = \mathbf{b}$ becomes the scalar equation

$$ax = b. \qquad 4.67$$

The scalar a^{-1} is the inverse of the scalar a, so we may multiply on both sides of Equation 4.67 to obtain the result

$$a^{-1}(ax) = a^{-1}b \qquad 4.68$$

$$1x = a^{-1}b.$$

We would like to generalize this simple idea to the matrix equation $\mathbf{A}\mathbf{x} = \mathbf{b}$ so that we can find an inverse of the matrix $\mathbf{A}$, called $\mathbf{A}^{-1}$, and write

$$\mathbf{A}^{-1}(\mathbf{A}\mathbf{x}) = \mathbf{A}^{-1}\mathbf{b} \qquad 4.69$$

$$\mathbf{Ix} = \mathbf{A}^{-1}\mathbf{b}.$$

In this equation the matrix $\mathbf{I}$ is the identity matrix

$$\mathbf{I} = \begin{bmatrix} 1 & 0 & 0 & \cdots & 0 \\ 0 & 1 & 0 & \cdots & 0 \\ 0 & 0 & 1 & \cdots & 0 \\ \vdots & \vdots & \vdots & \ddots & \vdots \\ 0 & 0 & 0 & \cdots & 1 \end{bmatrix}. \qquad 4.70$$

It is clear that the identity matrix $\mathbf{I}$ and the inverse of a matrix, $\mathbf{A}^{-1}$, play a key role in the solution of linear equations. Let's study them in more detail.

The Matrix Identity. When we multiply a scalar by 1, we get back that same scalar. For this reason, the number 1 is called the *multiplicative identity* element. This may seem trivial, but the generalization to matrices is more interesting. The question is, is there a matrix that, when it multiplies another matrix, does not change the other matrix? The answer is yes. The matrix is called the *identity matrix* and is denoted by $\mathbf{I}$. The identity matrix is always square, with whatever size is appropriate for the matrix multiplication. The identity matrix has 1's on its main diagonal and 0's everywhere else. For example,

$$\mathbf{I}_5 = \begin{bmatrix} 1 & 0 & 0 & 0 & 0 \\ 0 & 1 & 0 & 0 & 0 \\ 0 & 0 & 1 & 0 & 0 \\ 0 & 0 & 0 & 1 & 0 \\ 0 & 0 & 0 & 0 & 1 \end{bmatrix}. \qquad 4.71$$

The subscript 5 indicates the size. In terms of the unit coordinate vectors $\mathbf{e}_i$, we can also write the $n \times n$ identity matrix as

$$\mathbf{I}_n = \begin{bmatrix} | & | & & | \\ \mathbf{e}_1 & \mathbf{e}_2 & \cdots & \mathbf{e}_n \\ | & | & & | \end{bmatrix}. \qquad 4.72$$

For any matrix $\mathbf{A} \in \mathcal{R}^{n \times p}$, we have

$$\mathbf{A} = \mathbf{I}_n \mathbf{A}. \tag{4.73}$$

For $p = 1$, we obtain the following special form for any vector $\mathbf{x} \in \mathcal{R}^n$:

$$\mathbf{x} = \mathbf{I}_n \mathbf{x}. \tag{4.74}$$

This last equation can be written as the sum

$$\mathbf{x} = \sum_{i=1}^{n} x_i \mathbf{e}_i. \tag{4.75}$$

This is a special case of one of the results you proved in Problem 4.24. Figure 4.12 illustrates Equation 4.75 and shows how the vector $\mathbf{x}$ can be broken into component vectors $x_i \mathbf{e}_i$, each of which lies in the direction of one coordinate axis. The values of the x_i's are the coordinates of $\mathbf{x}$, and their magnitudes are also the lengths of the component vectors.

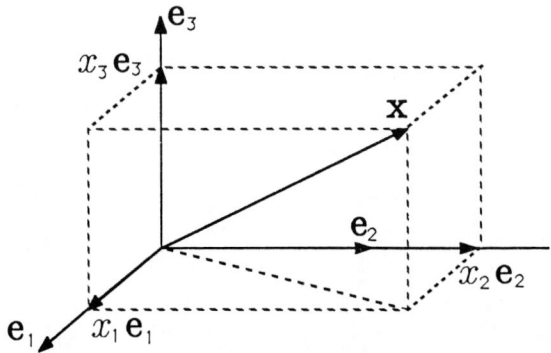

Figure 4.12: Breaking a Vector into Components

Problem 4.28   Use Equation 4.74 and the rules for matrix multiplication to show that $\mathbf{x} \in \mathcal{R}^n$ may also be written as

$$\mathbf{x} = \sum_{i=1}^{n} (\mathbf{x}, \mathbf{e}_i) \mathbf{e}_i. \ \blacksquare$$

This verifies Equation 4.27.

The Matrix Inverse. When the product of two numbers is 1 (the identity element), we say that they are inverses of each other, like 2 and 0.5. Likewise, we say that two *square* matrices are *inverses* of each other if their product is the identity matrix:

$$\mathbf{AB} = \mathbf{I} \quad \Longleftrightarrow \quad \mathbf{B} = \mathbf{A}^{-1}. \qquad 4.76$$

An interesting and useful result is implied by this definition. Take the first form of Equation 4.76 and multiply by $\mathbf{B}$ from the left:

$$\mathbf{AB} = \mathbf{I}$$
$$\Longrightarrow \quad \mathbf{B(AB)} = \mathbf{BI}$$
$$\Longrightarrow \quad \mathbf{(BA)B} = \mathbf{B}$$
$$\Longrightarrow \quad \mathbf{BA} = \mathbf{I}.$$

We emphasize that the inference made in the last step here is only valid when $\mathbf{B}$ and $\mathbf{A}$ are square matrices. The result means that, even though matrix multiplication is not commutative in general, we have a special case where it is commutative. If $\mathbf{A}$ and $\mathbf{B}$ are inverses of each other, then

$$\mathbf{AB} = \mathbf{BA} = \mathbf{I}. \qquad 4.77$$

Problem 4.29 Prove that the inverse of the 2×2 rotation matrix $\mathbf{R}(\theta)$ is $\mathbf{R}^T(\theta)$. ∎

Matrices that are not square do not have inverses. In fact, not all square matrices have inverses. So it becomes an important issue to determine which matrices do have inverses. If a matrix has an inverse, the inverse is unique. If a square matrix has no inverse, it is called a *singular* matrix. The *determinant* of a square matrix is a scalar computed from the numbers in the matrix. It tells us whether a matrix will have an inverse or not. A matrix is singular if and only if its determinant is zero.[3] In MATLAB, the notation

---

[3] It is not important now to understand how the determinant is defined and computed from the elements of a matrix. In your more advanced courses you will study the determinant in some detail.

`det(A)` is used to compute the determinant. Whenever the matrix $\mathbf{A}$ in Equation 4.65 is singular, it means one of two things about the system of equations represented: either the equations are inconsistent and there is no solution, or the equations are dependent and there are infinitely many solutions.

Solving $\mathbf{A}\mathbf{x} = \mathbf{b}$. Let's now study ways to solve the matrix equation $\mathbf{A}\mathbf{x} = \mathbf{b}$. We will assume that a unique solution for $\mathbf{x}$ exists. Thus a unique matrix $\mathbf{A}^{-1}$ exists with the property $\mathbf{A}^{-1}\mathbf{A} = \mathbf{I}$. The trick is to find it. Here is one procedure.

For convenience, rewrite the matrix equation $\mathbf{A}\mathbf{x} = \mathbf{b}$ as

$$\begin{bmatrix} \mathbf{A} & \mathbf{b} \end{bmatrix} \begin{bmatrix} \mathbf{x} \\ -1 \end{bmatrix} = \mathbf{0}. \tag{4.78}$$

The matrix $\begin{bmatrix} \mathbf{A} & \mathbf{b} \end{bmatrix} \in \mathcal{R}^{n \times (n+1)}$ is called the *augmented matrix* for the system of equations. The augmented matrix may be viewed as a systematic way of writing all the information necessary to solve the equations.

The advantage of Equation 4.78 is that we may premultiply both sides by any matrix $\mathbf{C}_1$ (of compatible size), and the right-hand side remains zero (although this is equivalent to multiplying on both sides of Equation 4.65, which some may prefer). We can repeat this operation as often as we like with matrices $\mathbf{C}_2$, $\mathbf{C}_3$, etc. The general result is

$$\begin{bmatrix} \mathbf{C}_m \cdots \mathbf{C}_2 \mathbf{C}_1 \mathbf{A} & \mathbf{C}_m \cdots \mathbf{C}_2 \mathbf{C}_1 \mathbf{b} \end{bmatrix} \begin{bmatrix} \mathbf{x} \\ -1 \end{bmatrix} = \mathbf{0}. \tag{4.79}$$

Now suppose that we have found a sequence of matrices $\mathbf{C}_1, \ldots, \mathbf{C}_m$ that transforms the matrix $\mathbf{A}$ to the identity matrix:

$$\mathbf{C}_m \cdots \mathbf{C}_2 \mathbf{C}_1 \mathbf{A} = \mathbf{I}. \tag{4.80}$$

The augmented matrix equation in 4.79 simplifies to

$$\begin{bmatrix} \mathbf{I} & \mathbf{C}_m \cdots \mathbf{C}_2 \mathbf{C}_1 \mathbf{b} \end{bmatrix} \begin{bmatrix} \mathbf{x} \\ -1 \end{bmatrix} = \mathbf{0}, \tag{4.81}$$

which can be multiplied out to reveal that the solution for $\mathbf{x}$ is

$$\mathbf{x} = \mathbf{C}_m \cdots \mathbf{C}_2 \mathbf{C}_1 \mathbf{b}. \tag{4.82}$$

The method may be summarized as follows:

(i) form the augmented matrix $[\mathbf{A}\ \mathbf{b}]$ for the system;

(ii) premultiply $[\mathbf{A}\ \mathbf{b}]$ by a sequence of selected matrices $\mathbf{C}_i$, designed to take $\mathbf{A}$ to $\mathbf{I}$; and

(iii) when $\mathbf{A}$ is transformed to $\mathbf{I}$, $\mathbf{b}$ will be transformed to $\mathbf{x}$, so the solution will appear as the last column of the transformed augmented matrix.

We may also conclude that the product of the matrices $\mathbf{C}_i$ must be the inverse of $\mathbf{A}$ since $\mathbf{A}^{-1}$ is the unique matrix for which $\mathbf{A}^{-1}\mathbf{A} = \mathbf{I}$. In solving for $\mathbf{x}$ by this method, we have found $\mathbf{A}^{-1}$ implicitly as the product $\mathbf{C}_m \cdots \mathbf{C}_2 \mathbf{C}_1$.

Example 4.9. Consider the equation

$$
\begin{bmatrix} 3 & 1 \\ 2 & 4 \end{bmatrix} \begin{bmatrix} x_1 \\ x_2 \end{bmatrix} = \begin{bmatrix} 6 \\ 5 \end{bmatrix}
$$
$$
\mathbf{A} \qquad \mathbf{x} \quad = \quad \mathbf{b}.
$$

The augmented matrix for this equation is

$$
[\mathbf{A}\ \mathbf{b}] = \begin{bmatrix} 3 & 1 & 6 \\ 2 & 4 & 5 \end{bmatrix}.
$$

Now if we could add $-2/3$ times the first row to the second row, we would get 0 in the lower left corner. This is the first step in transforming $\mathbf{A}$ to the identity $\mathbf{I}$. We can accomplish this row operation with the matrix

$$
\mathbf{C}_1 = \begin{bmatrix} 1 & 0 \\ -2/3 & 1 \end{bmatrix}
$$

$$
\mathbf{C}_1[\mathbf{A}\ \mathbf{b}] = \begin{bmatrix} 3 & 1 & 6 \\ 0 & 10/3 & 1 \end{bmatrix}.
$$

Now adding $-3/10$ times the new second row to the first row will introduce 0 in the $(1, 2)$ position, bringing us closer still to the identity. Thus

$$
\mathbf{C}_2 = \begin{bmatrix} 1 & -3/10 \\ 0 & 1 \end{bmatrix}
$$

$$\mathbf{C}_2\mathbf{C}_1\,[\mathbf{A}\ \mathbf{b}] = \begin{bmatrix} 3 & 0 & 57/10 \\ 0 & 10/3 & 1 \end{bmatrix}.$$

We now complete the transformation to identity by normalizing each row to get the needed 1's:

$$\mathbf{C}_3 = \begin{bmatrix} 1/3 & 0 \\ 0 & 3/10 \end{bmatrix}$$

$$\mathbf{C}_3\mathbf{C}_2\mathbf{C}_1\,[\mathbf{A}\ \mathbf{b}] = \begin{bmatrix} 1 & 0 & 19/10 \\ 0 & 1 & 3/10 \end{bmatrix}.$$

According to the last column, the solution is

$$\mathbf{x} = \begin{bmatrix} 19/10 \\ 3/10 \end{bmatrix}.$$

We note in passing that the inverse of $\mathbf{A}$ is the product of the $\mathbf{C}$'s, so

$$\mathbf{A}^{-1} = \mathbf{C}_3\mathbf{C}_2\mathbf{C}_1 = \begin{bmatrix} 1/3 & 0 \\ 0 & 3/10 \end{bmatrix} \begin{bmatrix} 1 & -3/10 \\ 0 & 1 \end{bmatrix} \begin{bmatrix} 1 & 0 \\ -2/3 & 1 \end{bmatrix}$$
$$= \begin{bmatrix} 1/3 & -1/10 \\ 0 & 3/10 \end{bmatrix} \begin{bmatrix} 1 & 0 \\ -2/3 & 1 \end{bmatrix} = \begin{bmatrix} 2/5 & -1/10 \\ -1/5 & 3/10 \end{bmatrix}. \ \square$$

The method we have just used, combined with a particular way of choosing the $\mathbf{C}_i$ matrices, is called *Gauss elimination*. Gauss elimination requires less computation than finding the inverse of $\mathbf{A}$ because only the effect of $\mathbf{A}^{-1}$ on the specific vector $\mathbf{b}$ is computed. MATLAB can solve for $\mathbf{x}$ by either method, as shown in Demo 4.2. For hand computations, we suggest choosing the $\mathbf{C}_i$ matrices so that $\mathbf{C}_1$ produces 0's everywhere below the diagonal in the first column, $\mathbf{C}_2$ produces 0's below the diagonal in the second column, and so on up to $\mathbf{C}_{n-1}$. Then $\mathbf{C}_n$ produces 0's above the diagonal in the nth column, $\mathbf{C}_{n+1}$ produces 0's above the diagonal in column $n-1$, etc. The last one, $\mathbf{C}_{2n-1}$, normalizes the diagonal elements to 1's. We have assumed for simplicity that no 0's on the diagonal will be encountered in hand computations.

Problem 4.30 Check that $\mathbf{A}^{-1}\mathbf{A} = \mathbf{I}$ in Example 4.9. ∎

Problem 4.31 Augment Equation 4.64 as in Equation 4.78 and use the technique of Gauss elimination to solve for **x**. ∎

Demo 4.2 (MATLAB). From the command level of MATLAB, solve the matrix equation of Example 4.1 by typing

```
≫ A = [1 1;0.3 -1.2]
≫ b = [85;0]
```

You have entered the matrices **A** and **b**, which describe the problem. You can now solve for **x** by finding the inverse of **A** and multiplying **b**:

```
≫ x = inv(A) * b
```

In this example the inverse is computed quickly because **A** is a small matrix. For a large (say, 30 × 30) matrix, the answer would take longer to compute, and the more efficient method of Gauss elimination would reduce waiting time. You can use Gauss elimination in MATLAB by typing

```
≫ x = A \ b
```

You should get the same answer. Now type the MATLAB code required to compute **Ax** and to show **Ax** − **b** = **0**. □

Problem 4.32 (MATLAB) Write the following system of equations as **Ax** = **b** and solve using MATLAB:

$$3(x_1 - x_3) + 2(x_2 - 1) - 6 = x_3$$

$$4x_3 = 7x_2 - 5$$

$$6(x_1 + x_2 + 2) = x_3 + 1.$$ ∎

4.9 Circuit Analysis

In this section we use the linear algebra we have developed to find the voltages and currents in a simple electrical circuit, such as the one shown in Figure 4.13. There are many reasons why this might be necessary; in this example we need to know the current flowing through the lamp to tell whether or not it will glow. Too little current will have no visible effect; too much current will cause the lamp to burn out. We will apply a few physical laws relating the voltages and currents in a circuit, turn these laws into systems of linear equations, and solve the equations for the voltages and currents.

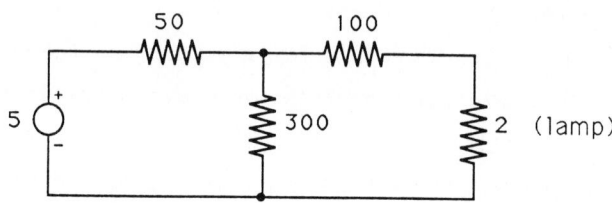

Figure 4.13: A Simple Electrical Circuit

Current, Voltage, and Resistance. We will use three physical quantities in our analysis of electrical circuits: current, voltage, and resistance. *Current* is the flow of electrical charge from one place to another. Electrons flowing through a wire or through some other electronic device comprise a current. *Voltage* is a difference in electric potential that makes electrons flow. Voltage is sometimes called *electromotive force* because it is like a "force" that moves electrons. *Resistance* is a property of the device through which the electron current flows. The lower the resistance of a device, the more easily current can flow through the device.

The analogy of water flowing through pipes can help you develop intuition about electrical circuits. In this analogy, electrical current corresponds to the flow rate of water. Voltage corresponds to the pressure that forces the water to flow, and resistance is the friction of flow. A small pipe would impede the flow of water more than would a large pipe, so the small pipe would correspond to a higher resistance. While this analogy can be helpful, keep in mind that electricity is not water. All analogies break down at some point.

We measure electrical current in *amperes*. The standard symbol for current is i, and the direction of positive flow is indicated by an arrow on the circuit diagram. The arrow is for reference only; if the true current is in the opposite direction, we get negative values for i. Because electrons are negatively charged, current is defined as flowing in the opposite direction as electron motion. But to reduce confusion, you should learn to think in terms of current rather than electron motion.

A point in a circuit where several devices are connected together is called a *node*. The conservation law for current says that "what flows in must flow out of a node," a principle known as *Kirchhoff's current law*. Kirchhoff's current law states that *the sum of all currents leaving a node is zero*. In this law, a current entering the node is considered to be a negative current leaving the node.

Voltage is measured in *volts* and is usually written as v (or e). Since voltage is a difference in potential between two points (nodes), we can show it on a circuit diagram with a + and a − sign to indicate which two nodes we are comparing and which one of the nodes is considered negative. As with current, the markings are for reference only and we may end up with a negative value of v.

In an electrical circuit, one node is usually chosen as a *reference node* and is considered to have a voltage of zero. Then the voltage at every other node is measured with respect to the reference node. This saves us the trouble of always specifying pairs of nodes for voltage measurements and marking + and − signs for each voltage. Other names for the reference node are *common* and *ground*.

A *constant voltage source* is a device that always forces the voltage between its two terminals to be a constant value. In Figure 4.13 the circle at the left represents a constant voltage source of 5 volts, so that the voltage at the upper (+) end is *always* exactly 5 volts higher than the voltage at the lower (−) end. A voltage source is something like a battery, but idealized. Real batteries do not maintain a constant output voltage under all conditions.

Resistance is measured in *ohms* and is denoted by R. A resistor is shown as a zig-zag line in circuit diagrams and labeled with the value of its resistance in ohms. In this chapter we will consider only devices whose resistance is positive and the same in both directions. *Ohm's law*, also called the *resistor law*, relates the voltage and current in a resistor. For the resistor shown in Figure 4.14, with reference directions assigned to v and i as shown, Ohm's law is

$$v = iR. \tag{4.83}$$

Note that current flows from + to − through the resistor.

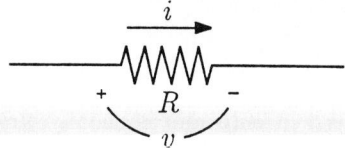

Figure 4.14: Ohm's Law

Example 4.10. Ohm's law and Kirchhoff's current law are the only principles we need to write equations that will allow us to find the voltages and currents in the resistive circuit of Figure 4.13. We begin by choosing a reference node and assigning variables to the voltages at every other node (with respect to the reference node). These choices are shown in Figure 4.15.

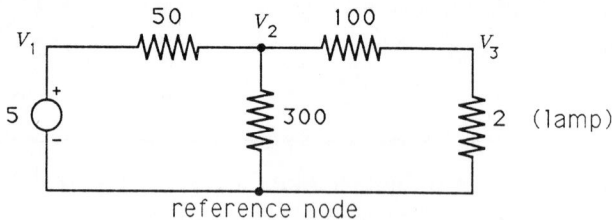

Figure 4.15: Assigning Node Voltages

The constant voltage source forces v_1 to be exactly 5 volts higher than the reference node. Thus

$$v_1 = 5. \tag{4.84}$$

Next we write equations by applying Kirchhoff's current law to each node in the circuit (except the reference node and v_1, whose voltages we already know). At the node labeled v_2 are three paths for leaving current. The current leaving through the 50 ohm resistor can be found by Ohm's law, where the voltage across that resistor is $v_2 - v_1$:

$$i_{50} = \frac{v}{R} = \frac{(v_2 - v_1)}{50}.$$

For current leaving through the 300 ohm resistor, the voltage is v_2. Pay careful attention to the sign; since we are interested in the current *leaving* the node labeled v_2, Figure 4.14 indicates that to apply Ohm's law we should take the voltage as $+v_2 -$ reference $= v_2 - 0 = v_2$. So

$$i_{300} = \frac{v_2}{300}.$$

For the 100 ohm resistor, we can write

$$i_{100} = \frac{(v_2 - v_3)}{100}.$$

According to Kirchhoff's current law, the sum of these three leaving currents is zero:

$$\frac{(v_2 - v_1)}{50} + \frac{v_2}{300} + \frac{(v_2 - v_3)}{100} = 0$$

$$\implies \quad 6(v_2 - v_1) + v_2 + 3(v_2 - v_3) = 0 \qquad\qquad 4.85$$

$$\implies \quad -6v_1 + 10v_2 - 3v_3 = 0.$$

Notice that when we wrote the equation for the node labeled v_2, the variable v_2 had a $+$ sign each time it occurred in the equation, while the others had a $-$ sign. This is always the case, and watching for it can help you avoid sign errors. Now we apply Kirchhoff's current law at the node labeled v_3 to get the equation

$$\frac{(v_3 - v_2)}{100} + \frac{v_3}{2} = 0$$

$$\implies \quad (v_3 - v_2) + 50v_3 = 0 \qquad\qquad 4.86$$

$$\implies \quad 0v_1 - 1v_2 + 51v_3 = 0.$$

Note that this time it is v_3 that always shows up with a $+$ sign.

Equations 4.84, 4.85, and 4.86 give us a system of three equations in the three unknown variables v_1, v_2, and v_3. We now write them in matrix form as

$$\begin{bmatrix} 1 & 0 & 0 \\ -6 & 10 & -3 \\ 0 & -1 & 51 \end{bmatrix} \begin{bmatrix} v_1 \\ v_2 \\ v_3 \end{bmatrix} = \begin{bmatrix} 5 \\ 0 \\ 0 \end{bmatrix}. \ \square \qquad 4.87$$

Problem 4.33 (MATLAB) Use MATLAB to solve Equation 4.87. You should find

$$v_1 = 5.0000 \text{ volts}$$
$$v_2 = 3.0178 \text{ volts}$$
$$v_3 = 0.0592 \text{ volt}.$$

What is the determinant of the coefficient matrix $\mathbf{A}$? Is the solution unique? ∎

We can determine the current flowing through the lamp from v_3 to ground in Example 4.10 by Ohm's law:

$$i = \frac{v}{R} = \frac{v_3}{2} = 0.0296 \text{ ampere.} \qquad 4.88$$

The visible effect will, of course, depend on the lamp. Let us assume that the specifications for our lamp indicate that 0.05 ampere or more is required before it will glow, and more than 0.075 ampere will cause it to burn out. In this case, our circuit would not make the lamp glow.

Problem 4.34 (MATLAB) Write and solve equations for the circuit in Figure 4.16. What are the voltages at the nodes labeled v_1 through v_4? What is the current labeled i_1? And i_2? ∎

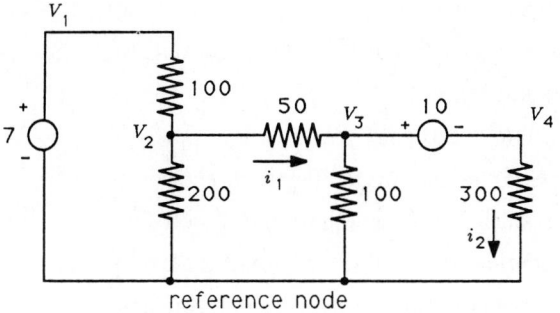

Figure 4.16: A Resistive Network

4.10 Numerical Experiment (Circuit Design)

Our analysis in Example 4.10 and Problem 4.33 indicates that not enough current will flow through the lamp to make it glow. We now wish to change the resistance of the 100 ohm resistor to a new value so that the lamp will glow. We replace 100 in the equations with an unknown resistance R. Equation 4.84 is unchanged, but Equation 4.85 becomes

$$\frac{(v_2 - v_1)}{50} + \frac{v_2}{300} + \frac{(v_2 - v_3)}{R} = 0$$

$$\implies \quad 6R(v_2 - v_1) + Rv_2 + 300(v_2 - v_3) = 0$$

$$\implies \quad -6Rv_1 + (7R + 300)v_2 - 300v_3 = 0.$$

Equation 4.86 becomes

$$\frac{(v_3 - v_2)}{R} + \frac{v_3}{2} = 0$$

$$\implies \quad 2(v_3 - v_2) + Rv_3 = 0$$

$$\implies \quad 0v_1 - 2v_2 + (R + 2)v_3 = 0.$$

The matrix form of these equations is

$$\begin{bmatrix} 1 & 0 & 0 \\ -6R & 7R + 300 & -300 \\ 0 & -2 & R + 2 \end{bmatrix} \begin{bmatrix} v_1 \\ v_2 \\ v_3 \end{bmatrix} = \begin{bmatrix} 5 \\ 0 \\ 0 \end{bmatrix}. \qquad 4.89$$

Write a MATLAB function file called `builda` to accept R as an input and return the matrix $\mathbf{A}$ in Equation 4.89 as an output. The first line of your function file should be

    function A = builda(R);

Now choose several values for R. For each choice, use your function `builda` and solve the resulting matrix equation $\mathbf{Av} = \mathbf{b}$ for the voltages. Each time you choose a different R to build a different matrix $\mathbf{A}$, check the determinant of $\mathbf{A}$ to be sure the equations have a unique solution:

    ≫ det(A)

Make a table of R and the corresponding values of v_3:

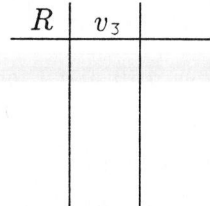

Now add a column to your table for the current through the lamp $i = v_3/2$. Add rows to your table until you have found a value of R for which the lamp will glow. (i needs to be between 0.05 and 0.075 ampere.)

5

Vector Graphics

Notes to Teachers and Students:

In this chapter we introduce matrix data structures that may be used to represent two- and three-dimensional images. The demonstration program Demo 5.1 (MATLAB) shows students how to create a function file for creating images from these data structures. We then show how to use matrix transformations for translating, scaling, and rotating images. Projections are used to project three-dimensional images onto two-dimensional planes placed at arbitrary locations. It is precisely such projections that we use to get perspective drawings on a two-dimensional surface of three-dimensional objects. The numerical experiment in Section 5.8 encourages students to manipulate a star field and view it from several points in space.

Once again we consider certain problems essential to the chapter development. For this chapter be sure not to miss Problems 5.2, 5.5, 5.6, 5.11, and 5.13.

5.1 Introduction

Pictures play a vital role in human communication, in robotic manufacturing, and in digital imaging. In a typical application of digital imaging, a CCD camera records a digital picture frame that is read into the memory of a digital computer. The digital computer then manipulates this frame (or array) of data in order to crop, enlarge or reduce, enhance or smooth, translate or rotate the original picture. These procedures are called *digital picture processing* or *computer graphics*. When a sequence of picture frames is processed and displayed at video frame rates (30 frames per second), then we have an animated picture.

Acknowledgments: *Fundamentals of Interactive Computer Graphics* by J. D. Foley and A. Van Dam, ©1982 Addison-Wesley Publishing Company, Inc., Reading, Massachusetts, was used extensively as a reference book during development of this chapter. Star locations were obtained from the shareware program "Deep Space" by David Chandler, who obtained them from the "Skymap" database of the National Space Science Data Center.

In this chapter we use the linear algebra we developed in Chapter 4 to develop a rudimentary set of tools for doing computer graphics on line drawings. We begin with an example: the rotation of a single point in the (x, y) plane.

Example 5.1. Point P has coordinates $(3, 1)$ in the (x, y) plane as shown in Figure 5.1. Find the coordinates of the point P', which is rotated $\frac{\pi}{6}$ radians from P.

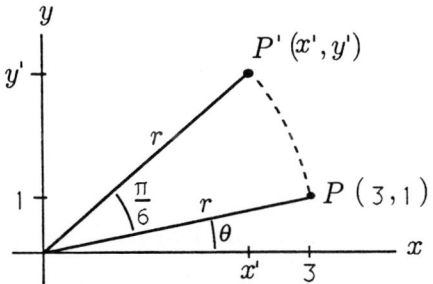

Figure 5.1: Rotating a Single Point in the $(\mathbf{x}, \mathbf{y})$ Plane

To solve this problem, we can begin by converting the point P from rectangular coordinates to polar coordinates. We have

$$r = \sqrt{x^2 + y^2} = \sqrt{10} \qquad\qquad 5.1$$

$$\theta = \tan^{-1}\left(\frac{y}{x}\right) \approx 0.3217 \text{ radian.}$$

The rotated point P' has the same radius r, and its angle is $\theta + \frac{\pi}{6}$. We now convert back to rectangular coordinates to find x' and y' for point P':

$$x' = r \cos\left(\theta + \frac{\pi}{6}\right) \approx \sqrt{10}\cos(0.8453) \approx 2.10 \qquad (5.2)$$

$$y' = r \sin\left(\theta + \frac{\pi}{6}\right) \approx \sqrt{10}\sin(0.8453) \approx 2.37.$$

So the rotated point P' has coordinates $(2.10, 2.37)$. $\square$

Now imagine trying to rotate the graphical image of some complex object like an airplane. You could try to rotate all 10,000 (or so) points in the same way as the single point was just rotated. However, a much easier way to rotate all the points together is provided by linear algebra. In fact, with a single linear algebraic operation we can rotate and scale an entire object and project it from three dimensions to two for display on a flat screen or sheet of paper.

In this chapter we study *vector graphics*, a linear algebraic method of storing and manipulating computer images. Vector graphics is especially suited to moving, rotating, and scaling (enlarging and reducing) images and objects within images. Cropping is often necessary too, although it is a little more difficult with vector graphics. Vector graphics also allows us to store objects in three dimensions and then view the objects from various locations in space by using projections.

In vector graphics, pictures are drawn from straight lines.[1] A curve can be approximated as closely as desired by a series of short, straight lines. Clearly some pictures are better suited to representation by straight lines than are others. For example, we can achieve a fairly good representation of a building or an airplane in vector graphics, while a photograph of a forest would be extremely difficult to convert to straight lines. Many computer-aided design (CAD) programs use vector graphics to manipulate mechanical drawings.

When the time comes to actually display a vector graphics image, it may be necessary to alter the representation to match the display device. Personal computer display screens are divided into thousands of tiny rectangles called *picture elements*, or *pixels*. Each pixel is either off (black) or on (perhaps with variable intensity and/or color). With a CRT display, the electron beam scans the rows of pixels in a raster pattern. To draw a line on a pixel display device, we must first convert the line into a list of pixels to be illuminated. Dot matrix and laser printers are also pixel display devices, while pen plotters and a few specialized CRT devices can display vector graphics

---

[1] It is possible to extend these techniques to deal with some types of curves, but we will consider only straight lines for the sake of simplicity.

directly. We will let MATLAB do the conversion to pixels and automatically handle cropping when necessary.

We begin our study of vector graphics by representing each point in an image by a vector. These vectors are arranged side-by-side into a matrix $\mathbf{G}$ containing all the points in the image. Other matrices will be used as operators to perform the desired transformations on the image points. For example, we will find a matrix $\mathbf{R}$, which functions as a rotation: the matrix product $\mathbf{RG}$ represents a rotated version of the original image $\mathbf{G}$.

5.2 Two-Dimensional Image Representation

Point Matrix. To represent a straight-line image in computer memory, we must store a list of all the endpoints of the line segments that comprise the image. If the point $P_i = (x_i, y_i)$ is such an endpoint, we write it as the column vector

$$\mathbf{p}_i = \begin{bmatrix} x_i \\ y_i \end{bmatrix}. \qquad 5.3$$

Suppose there are n such endpoints in the entire image. Each point is included only once, even if several lines end at the same point. We can arrange the vectors $\mathbf{p}_i$ into a point matrix:

$$\begin{aligned} \mathbf{G} &= \begin{bmatrix} \mathbf{p}_1 & \mathbf{p}_2 & \mathbf{p}_3 & \cdots & \mathbf{p}_n \end{bmatrix} \\ &= \begin{bmatrix} x_1 & x_2 & x_3 & \cdots & x_n \\ y_1 & y_2 & y_3 & \cdots & y_n \end{bmatrix}. \end{aligned} \qquad 5.4$$

We then store the point matrix $\mathbf{G} \in \mathcal{R}^{2 \times n}$ as a two-dimensional array in computer memory.

Example 5.2. Consider the list of points

$$\begin{aligned} P_1 &= (0, 0) & (5.5) \\ P_2 &= (-1.5, 5) \\ P_3 &= (4, 2.3) \\ P_4 &= (4, -1). \end{aligned}$$

The corresponding point matrix is

$$\mathbf{G} = \begin{bmatrix} 0 & -1.5 & 4 & 4 \\ 0 & 5 & 2.3 & -1 \end{bmatrix}. \quad \square \qquad\qquad 5.6$$

Line Matrix. The next thing we need to know is which pairs of points to connect with lines. To store this information for m lines, we will use a line matrix, $\mathbf{H} \in \mathcal{R}^{2 \times m}$. The line matrix *does not* store line locations directly. Rather, it contains references to the points stored in $\mathbf{G}$. To indicate a line between points $\mathbf{p}_i$ and $\mathbf{p}_j$, we store the indices i and j as a pair. For the kth line in the image, we have the pair

$$\mathbf{h}_k = \begin{bmatrix} i_k \\ j_k \end{bmatrix}. \qquad\qquad 5.7$$

The order of i and j does not really matter since a line from $\mathbf{p}_i$ to $\mathbf{p}_j$ is the same as a line from $\mathbf{p}_j$ to $\mathbf{p}_i$. Next we collect all the lines $\mathbf{h}_k$ into a line matrix $\mathbf{H}$:

$$\mathbf{H} = \begin{bmatrix} i_1 & i_2 & i_3 & \cdots & i_m \\ j_1 & j_2 & j_3 & \cdots & j_m \end{bmatrix}. \qquad\qquad 5.8$$

All the numbers in the line matrix $\mathbf{H}$ will be positive integers since they point to columns of $\mathbf{G}$. To find the actual endpoints of a line, we look at columns i and j of the point matrix $\mathbf{G}$.

Example 5.3. To specify line segments connecting the four points of Example 5.2 into a quadrilateral, we use the line matrix

$$\mathbf{H}_1 = \begin{bmatrix} 1 & 2 & 3 & 4 \\ 2 & 3 & 4 & 1 \end{bmatrix}. \qquad\qquad 5.9$$

Alternatively, we can specify line segments to form a triangle from the first three points plus a line from P_3 to P_4:

$$\mathbf{H}_2 = \begin{bmatrix} 1 & 2 & 3 & 3 \\ 2 & 3 & 1 & 4 \end{bmatrix}. \qquad\qquad 5.10$$

Figure 5.2 shows the points $\mathbf{G}$ connected first by $\mathbf{H}_1$ and then by $\mathbf{H}_2$. $\square$

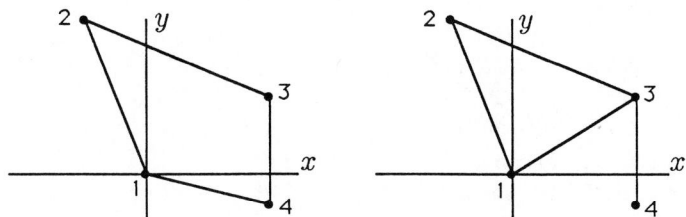

Figure 5.2: Two Sets of Lines

Demo 5.1 (MATLAB). Use your editor to enter the following MATLAB function file. Save it as `vgraph1.m`.

```
function vgraph1(points,lines);
% vgraph1(points,lines) plots the points as *'s and
% connects the points with specified lines.  The points
% matrix should be 2xN, and the lines matrix should be 2xM.
% The field of view is preset to (-50,50) on both axes.
%
% Written by Richard T. Behrens, October 1989.
%
m = length(lines);               % find the number of
                                 % lines.
axis([-50 50 -50 50])            % set the axis scales
axis('square')
plot(points(1,:),points(2,:),'*')    % plot the points as *
hold on                          % keep the points...
for i = 1:m                      % while plotting the
                                 % lines
   plot([points(1,lines(1,i)) points(1,lines(2,i))], ..
        [points(2,lines(1,i)) points(2,lines(2,i))],'-')

end

hold off
```

After you have saved the function file, run MATLAB and type the following to enter the point and line matrices. (We begin with the transposes of the matrices to make them easier to enter.)

```
≫ G = [
     0.6052  -0.4728;
    -0.4366   3.5555;
    -2.6644   7.9629;
    -7.2541  10.7547;
   -12.5091  11.5633;
   -12.5895  15.1372;
    -6.5602  13.7536;
   -31.2815  -7.7994;
   -38.6185  -9.9874;
   -44.0593  -1.1537;
   -38.8315   2.5452;
   -39.4017   9.4595;
   -39.3192  15.0932;
   -45.9561  23.4158]

≫ G = G'

≫ H = [
    1    2;
    2    3;
    3    4;
    4    5;
    4    7;
    5    6;
    8    9;
    9   10;
   10   11;
   11   12;
   12   13;
   13   14]

≫ H = H'
```

At this point you should use MATLAB's "save" command to save these matrices to a disk file. Type

```
≫ save dippers
```

After you have saved the matrices, use the function **VGRAPH1** to draw the image by typing

```
≫ vgraph1(G,H)   □
```

The advantage of storing points and lines separately is that an object can be moved and scaled by operating only on the point matrix $\mathbf{G}$. The line information in $\mathbf{H}$ remains the same since the same pairs of points are connected no matter where we put the points themselves.

Surfaces and Objects. To describe a surface in three dimensions is a fairly complex task, especially if the surface is curved. For this reason, we will be satisfied with points and lines, sometimes visualizing flat surfaces based on the lines. On the other hand, it is a fairly simple matter to group the points and lines into distinct *objects*. We can define an object matrix $\mathbf{K}$ with one column for each object giving the ranges of points and lines associated with that object. Each column is defined as

$$\mathbf{k}_i = \begin{bmatrix} \text{first point} \\ \text{last point} \\ \text{first line} \\ \text{last line} \end{bmatrix}.$$

As with the line matrix $\mathbf{H}$, the elements of $\mathbf{K}$ are integers.

Example 5.4. Consider again Demo 5.1. We could group the points in $\mathbf{G}$ and the lines in $\mathbf{H}$ into two objects with the matrix

$$\mathbf{K} = \begin{bmatrix} 1 & 8 \\ 7 & 14 \\ 1 & 7 \\ 6 & 12 \end{bmatrix}. \tag{5.11}$$

The first column of $\mathbf{K}$ specifies that the first object (Ursa Minor) is made up of points 1 through 7 and lines 1 through 6, and the second column of $\mathbf{K}$ defines the second object (Ursa Major) as points 8 through 14 and lines 7 through 12. □

5.3 Two-Dimensional Image Transformations

We now turn our attention to operating on the point matrix $\mathbf{G}$ to produce the desired transformations. We will consider (i) rotation, (ii) scaling, and (iii) translation (moving) of objects. Rotation and scaling are done by matrix multiplication with a square transformation matrix $\mathbf{A}$. If we call the transformed point matrix $\mathbf{G}_{\text{new}}$, we have

$$[\quad \mathbf{G}_{\text{new}} \quad] = [\mathbf{A}] [\quad \mathbf{G} \quad]. \qquad 5.12$$

We call $\mathbf{A}$ a *matrix operator* because it "operates" on $\mathbf{G}$ through matrix multiplication. In contrast, translation must be done by matrix addition.

In a later section you will see that it is advantageous to perform all operations by matrix operators and that we can modify our image representation to allow translation to be done with a matrix operator like rotation and scaling. We will call the modified representation *homogeneous coordinates.*

Rotation. We saw in Chapter 4 that the matrix that rotates points by an angle θ is

$$\mathbf{A} = \mathbf{R}(\theta) = \begin{bmatrix} \cos\theta & -\sin\theta \\ \sin\theta & \cos\theta \end{bmatrix}. \qquad 5.13$$

When applied to the point matrix $\mathbf{G}$, this matrix operator rotates each point by the angle θ, regardless of the number of points.

Example 5.5. We can use the rotation matrix to do the single point rotation of Example 5.1. We have a point matrix consisting of only the point $(3, 1)$:

$$\mathbf{G} = \begin{bmatrix} 3 \\ 1 \end{bmatrix}. \qquad 5.14$$

The necessary transformation matrix is $\mathbf{R}(\theta)$ with $\theta = \frac{\pi}{6}$. Then the rotated point is given by

$$\mathbf{G}_{\text{new}} = \mathbf{R}\left(\frac{\pi}{6}\right)\mathbf{G} = \begin{bmatrix} \cos\left(\dfrac{\pi}{6}\right) & -\sin\left(\dfrac{\pi}{6}\right) \\ \sin\left(\dfrac{\pi}{6}\right) & \cos\left(\dfrac{\pi}{6}\right) \end{bmatrix} \begin{bmatrix} 3 \\ 1 \end{bmatrix} \approx \begin{bmatrix} 2.10 \\ 2.37 \end{bmatrix}. \quad \square \qquad 5.15$$

Scaling. An object can be enlarged or reduced in each dimension independently. The matrix operator that scales an image by a factor of s_x along the x-axis and s_y along the y-axis is

$$\mathbf{A} = \mathbf{S}(s_x, s_y) = \begin{bmatrix} s_x & 0 \\ 0 & s_y \end{bmatrix}.$$ 5.16

Most often we take $s_x = s_y$ to scale an image by the same amount in both dimensions.

Problem 5.1   Write out the following matrices. Simplify and give numerical answers to two decimal places:

(a) $\mathbf{R}(\frac{\pi}{2})$;

(b) $\mathbf{S}(3, 2)$;

(c) $\mathbf{R}(-\frac{\pi}{4})$;

(d) $\mathbf{S}(-1, 1)$. ∎

Problem 5.2   (Reflections) What does $\mathbf{S}(-1, 1)$ do? $\mathbf{S}(1, -1)$? $\mathbf{S}(-1, -1)$? $\mathbf{S}(1, 1)$? ∎

Problem 5.3   Given $\mathbf{G} = \begin{bmatrix} 0 & -1.5 & 4 & 4 \\ 0 & 5 & 2.3 & -1 \end{bmatrix}$ and $\theta = \frac{\pi}{3}$, find $\mathbf{G}_{\text{new}} = \mathbf{R}(\theta)\mathbf{G}$. Give numerical answers to two decimal places. ∎

Problem 5.4   Apply each of the transformations in Problems 5.1 and 5.2 to the image

$$\mathbf{G} = \begin{bmatrix} 1 & 1 & 2 & 2 \\ 1 & 2 & 2 & 1 \end{bmatrix}; \qquad \mathbf{H} = \begin{bmatrix} 1 & 2 & 3 & 4 \\ 2 & 3 & 4 & 1 \end{bmatrix}.$$

Sketch the original image and each transformation of it. ∎

Translation. An object can be moved by adding a constant vector $\mathbf{b}$ to every point in the object. For example, $\mathbf{b} = \begin{bmatrix} 20 \\ -5 \end{bmatrix}$ will move an object 20 units to the right and 5 units down. We can write this in terms of the point matrix as

$$\mathbf{G}_{new} = \mathbf{G} + \mathbf{b}\,\mathbf{1}^T \qquad 5.17$$

where $\mathbf{1}$ (read "the one-vector") is a vector of n 1's:

$$\mathbf{1} = \begin{bmatrix} 1 \\ 1 \\ \vdots \\ 1 \end{bmatrix}. \qquad 5.18$$

In MATLAB, $\mathbf{1}$ may be obtained by ones(n,1). The outer product of $\mathbf{b}$ with $\mathbf{1}$ in Equation 5.17 simply serves to make n copies of $\mathbf{b}$ so that one copy can be added to each point in $\mathbf{G}$.

5.4 Composition of Transformations

Often we will want to perform several operations on an object before we display the result. For example, suppose we want to rotate by $\frac{\pi}{3}$ and reduce to $\frac{1}{2}$ size in each dimension:

$$\mathbf{G}_1 = \mathbf{R}\!\left(\frac{\pi}{3}\right)\mathbf{G} \qquad 5.19$$
$$\mathbf{G}_{new} = \mathbf{S}\!\left(\tfrac{1}{2},\tfrac{1}{2}\right)\mathbf{G}_1.$$

If there are n points in the matrix $\mathbf{G}$, it will require $4n$ multiplications to perform each of these operations, for a total of $8n$ multiplications. However, we can save some multiplications by noting that

$$\mathbf{G}_{new} = \mathbf{S}\!\left(\tfrac{1}{2},\tfrac{1}{2}\right)\left[\mathbf{R}\!\left(\frac{\pi}{3}\right)\mathbf{G}\right] = \mathbf{A}\mathbf{G}$$

where

$$\mathbf{A} = \mathbf{S}\left(\tfrac{1}{2}, \tfrac{1}{2}\right)\mathbf{R}\left(\frac{\pi}{3}\right)$$

$$= \begin{bmatrix} \tfrac{1}{2}\cos\left(\dfrac{\pi}{3}\right) & -\tfrac{1}{2}\sin\left(\dfrac{\pi}{3}\right) \\ \tfrac{1}{2}\sin\left(\dfrac{\pi}{3}\right) & \tfrac{1}{2}\cos\left(\dfrac{\pi}{3}\right) \end{bmatrix}.$$

In other words, we take advantage of the fact that matrix multiplication is associative to combine $\mathbf{S}$ and $\mathbf{R}$ into a single operation $\mathbf{A}$, which requires only 8 multiplications. Then we operate on $\mathbf{G}$ with $\mathbf{A}$, which requires $4n$ multiplications. By "composing" the two operations, we have reduced the total from $8n$ to $4n + 8$ multiplications. Furthermore, we can now build operators with complex actions by combining simple actions.

Example 5.6. We can build an operator that stretches objects along a diagonal line by composing scaling and rotation. We must

(i) rotate the diagonal line to the x-axis with $\mathbf{R}(-\theta)$;

(ii) scale with $\mathbf{S}(s, 1)$; and

(iii) rotate back to the original orientation with $\mathbf{R}(\theta)$.

Figure 5.3 shows a square being stretched along a 45° line. The composite operator that performs this directional stretching is

$$\mathbf{A}(\theta, s) = \mathbf{R}(\theta)\mathbf{S}(s, 1)\mathbf{R}(-\theta)$$

$$= \begin{bmatrix} \cos\theta & -\sin\theta \\ \sin\theta & \cos\theta \end{bmatrix} \begin{bmatrix} s & 0 \\ 0 & 1 \end{bmatrix} \begin{bmatrix} \cos\theta & \sin\theta \\ -\sin\theta & \cos\theta \end{bmatrix}$$

$$= \begin{bmatrix} s\cos^2\theta + \sin^2\theta & (s-1)\sin\theta\cos\theta \\ (s-1)\sin\theta\cos\theta & \cos^2\theta + s\sin^2\theta \end{bmatrix}.$$

Note that the rightmost operator in a product of operators is applied first. □

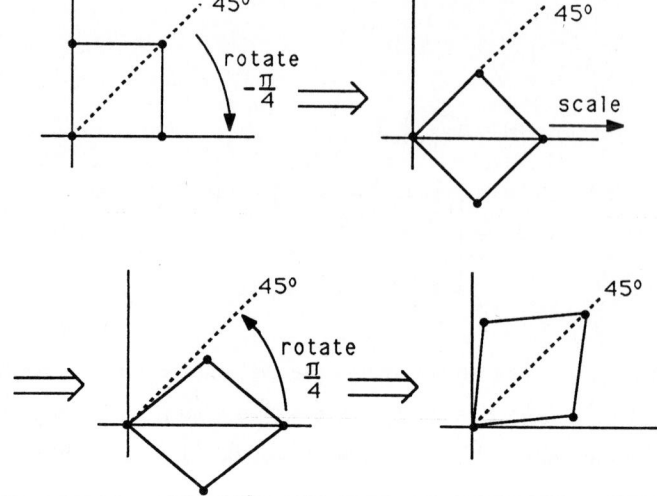

Figure 5.3: Rotating and Scaling for Directional Stretching

5.5 Homogeneous Coordinates

In the previous section we were able to combine rotation and scaling into a single composite operation by matrix multiplication. Unfortunately, translation cannot yet be included in the composite operator since we do it by addition rather than by multiplication.

Example 5.7. Suppose we wish to rotate the image $\mathbf{G}$ by $\frac{\pi}{3}$ about the point $(-10, 10)$. Our rotation matrix $\mathbf{R}(\theta)$ always rotates about the origin, so we must combine three transformations to accomplish this:

(i) translate the point $(-10, 10)$ to the origin;

(ii) rotate $\frac{\pi}{3}$ radians about the origin; and

(iii) translate the origin back to $(-10, 10)$.

For step (i), we have $\mathbf{b}_0 = \begin{bmatrix} 10 \\ -10 \end{bmatrix}$ and

$$\mathbf{G}_1 = \mathbf{G} + \mathbf{b}_0 \mathbf{1}^T.$$

For step (ii),

$$\begin{aligned}
\mathbf{G}_2 &= \mathbf{R}\!\left(\frac{\pi}{3}\right)\mathbf{G}_1 \\
&= \mathbf{R}\!\left(\frac{\pi}{3}\right)[\mathbf{G} + \mathbf{b}_0\mathbf{1}^T] \\
&= \mathbf{R}\!\left(\frac{\pi}{3}\right)\mathbf{G} + \mathbf{R}\!\left(\frac{\pi}{3}\right)\mathbf{b}_0\mathbf{1}^T.
\end{aligned}$$

For step (iii), we can use $-\mathbf{b}_0$ from step (i):

$$\begin{aligned}
\mathbf{G}_{\mathrm{new}} &= \mathbf{G}_2 - \mathbf{b}_0\mathbf{1}^T \\
&= \mathbf{R}\!\left(\frac{\pi}{3}\right)\mathbf{G} + \mathbf{R}\!\left(\frac{\pi}{3}\right)\mathbf{b}_0\mathbf{1}^T - \mathbf{b}_0\mathbf{1}^T \\
&= \mathbf{R}\!\left(\frac{\pi}{3}\right)\mathbf{G} + \left[\left(\mathbf{R}\!\left(\frac{\pi}{3}\right) - \mathbf{I}\right)\mathbf{b}_0\right]\mathbf{1}^T. \quad \square
\end{aligned}$$

In this example we were unable to find a single matrix operator $\mathbf{A}$ to do the entire job. The total transformation took the form

$$\mathbf{G}_{\mathrm{new}} = \mathbf{A}\mathbf{G} + \mathbf{b}\mathbf{1}^T.$$

This is called an *affine transformation* because it involves both multiplication by $\mathbf{A}$ and addition of a constant matrix. This is in contrast to the more desirable *linear transformation*, which involves only multiplication by $\mathbf{A}$.

We will now move toward a modified representation of the image and the operators by rewriting the last equation as

$$\mathbf{G}_{\mathrm{new}} = \begin{bmatrix} \mathbf{A} & \mathbf{b} \end{bmatrix} \begin{bmatrix} \mathbf{G} \\ \mathbf{1}^T \end{bmatrix} \qquad 5.20$$

where in the example we had $\mathbf{A} = \mathbf{R}\!\left(\frac{\pi}{3}\right)$ and $\mathbf{b} = \left(\mathbf{R}\!\left(\frac{\pi}{3}\right) - \mathbf{I}\right)\mathbf{b}_0$.

Problem 5.5 Show that, for any matrices $\mathbf{A}, \mathbf{B}, \mathbf{C}, \mathbf{D}$ of compatible sizes,

$$\mathbf{A}\mathbf{B} + \mathbf{C}\mathbf{D} = \begin{bmatrix} \mathbf{A} & \mathbf{C} \end{bmatrix} \begin{bmatrix} \mathbf{B} \\ \mathbf{D} \end{bmatrix}. \quad \blacksquare$$

The matrix $\begin{bmatrix} \mathbf{G} \\ \mathbf{1}^T \end{bmatrix}$ looks like

$$\begin{bmatrix} x_1 & x_2 & \cdots & x_n \\ y_1 & y_2 & \cdots & y_n \\ 1 & 1 & \cdots & 1 \end{bmatrix},$$

and the points $(x_i, y_i, 1)$ are called *homogeneous coordinates*. We can modify Equation 5.20 so that the new point matrix is also in homogeneous coordinates:

$$\begin{bmatrix} \mathbf{G}_{new} \\ \mathbf{1}^T \end{bmatrix} = \begin{bmatrix} \mathbf{A} & \mathbf{b} \\ \mathbf{0}^T & 1 \end{bmatrix} \begin{bmatrix} \mathbf{G} \\ \mathbf{1}^T \end{bmatrix}.$$

In the new representation, each point in the image has a third coordinate, which is always a 1. The homogeneous transformation is a 3×3 matrix,

$$\mathbf{A}_h = \begin{bmatrix} \mathbf{A} & \mathbf{b} \\ \mathbf{0}^T & 1 \end{bmatrix},$$

which is capable of translation, rotation, and scaling all by matrix multiplication. Thus, using homogeneous coordinates, we can build composite transformations that include translation.

In homogeneous coordinates, we have

$$\mathbf{R}(\theta) = \begin{bmatrix} \cos\theta & -\sin\theta & 0 \\ \sin\theta & \cos\theta & 0 \\ 0 & 0 & 1 \end{bmatrix} \qquad 5.21$$

$$\mathbf{S}(s_x, s_y) = \begin{bmatrix} s_x & 0 & 0 \\ 0 & s_y & 0 \\ 0 & 0 & 1 \end{bmatrix}$$

$$\mathbf{T}(t_x, t_y) = \begin{bmatrix} 1 & 0 & t_x \\ 0 & 1 & t_y \\ 0 & 0 & 1 \end{bmatrix}.$$

Example 5.8. The composite transformation to triple the size of an image and *then* move it 2 units to the left is

$$\mathbf{A} = \begin{bmatrix} 1 & 0 & -2 \\ 0 & 1 & 0 \\ 0 & 0 & 1 \end{bmatrix} \begin{bmatrix} 3 & 0 & 0 \\ 0 & 3 & 0 \\ 0 & 0 & 1 \end{bmatrix} = \begin{bmatrix} 3 & 0 & -2 \\ 0 & 3 & 0 \\ 0 & 0 & 1 \end{bmatrix}.$$

On the other hand, the composite transformation to move an image 2 units to the left and *then* triple its size is

$$\mathbf{B} = \begin{bmatrix} 3 & 0 & 0 \\ 0 & 3 & 0 \\ 0 & 0 & 1 \end{bmatrix} \begin{bmatrix} 1 & 0 & -2 \\ 0 & 1 & 0 \\ 0 & 0 & 1 \end{bmatrix} = \begin{bmatrix} 3 & 0 & -6 \\ 0 & 3 & 0 \\ 0 & 0 & 1 \end{bmatrix}.$$

In the latter case, the distance of the translation is also tripled. $\square$

Problem 5.6   Find a single composite transformation in homogeneous coordinates that rotates an image by angle θ about point (x_i, y_i) as in Example 5.7. ∎

5.6 Three-Dimensional Homogeneous Coordinates

We now consider the storage and manipulation of three-dimensional objects. We will continue to use homogeneous coordinates so that translation can be included in composite operators. Homogeneous coordinates in three dimensions will also allow us to do perspective projections so that we can view a three-dimensional object from any point in space.

Image Representation. The three-dimensional form of the point matrix in homogeneous coordinates is

$$\mathbf{G} = \begin{bmatrix} x_1 & x_2 & x_3 & \cdots & x_n \\ y_1 & y_2 & y_3 & \cdots & y_n \\ z_1 & z_2 & z_3 & \cdots & z_n \\ 1 & 1 & 1 & \cdots & 1 \end{bmatrix} \in \mathcal{R}^{4 \times n}.$$

The line matrix $\mathbf{H}$ is exactly as before, pointing to pairs of columns in $\mathbf{G}$ to connect with lines. Any other grouping matrices for objects, etc., are also unchanged.

Image manipulations are done by a 4×4 matrix $\mathbf{A}$. To ensure that the fourth coordinate remains a 1, the operator $\mathbf{A}$ must have the structure

$$\mathbf{A} = \begin{bmatrix} a_{11} & a_{12} & a_{13} & a_{14} \\ a_{21} & a_{22} & a_{23} & a_{24} \\ a_{31} & a_{32} & a_{33} & a_{34} \\ 0 & 0 & 0 & 1 \end{bmatrix}. \qquad 5.22$$

The new image has point matrix

$$\mathbf{G}_{\text{new}} = \mathbf{A}\mathbf{G}.$$

Problem 5.7 If the coordinates of the ith point in **G** are $(x_i, y_i, z_i, 1)$, what are the coordinates of the ith point in $\mathbf{G}_{new} = \mathbf{AG}$ when **A** is as given in Equation 5.22? ∎

Problem 5.8 Write down the point matrix **G** for the unit cube (the cube with sides of length 1, with one corner at the origin and extending in the positive direction along each axis). Draw a sketch of the cube, numbering the vertices according to their order in your point matrix. Now write down the line matrix **H** to complete the representation of the cube. ∎

Left- and Right-Handed Coordinate Systems. In this book we work exclusively with *right-handed* coordinate systems. However, it is worth pointing out that there are two ways to arrange the axes in three dimensions. Figure 5.4(a) shows the usual *right-handed* coordinates, and the *left-handed* variation is shown in Figure 5.4(b). All possible choices of labels x, y, and z for the three coordinate axes can be rotated to fit one of these two figures, but no rotation will go from one to the other.

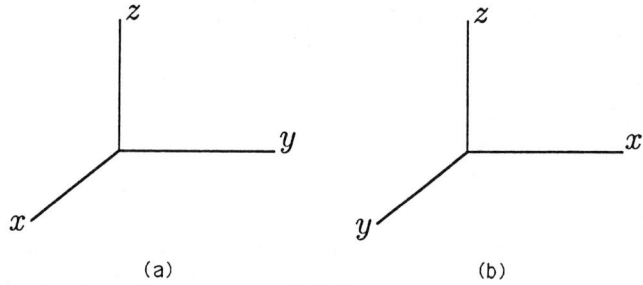

(a)   (b)

Figure 5.4: Three-Dimensional Coordinate Systems; (a) Right-handed, and (b) Left-handed

Be careful to sketch a right-handed coordinate system when you are solving problems in this chapter. Some answers will not be the same for a left-handed system.

Image Transformation. Three-dimensional operations are a little more complicated than their two-dimensional counterparts. For scaling and translation we now have three independent directions, so we generalize the operators of Equation 5.21 as

$$\mathbf{S}(s_x, s_y, s_z) = \begin{bmatrix} s_x & 0 & 0 & 0 \\ 0 & s_y & 0 & 0 \\ 0 & 0 & s_z & 0 \\ 0 & 0 & 0 & 1 \end{bmatrix} \qquad 5.23$$

$$\mathbf{T}(t_x, t_y, t_z) = \begin{bmatrix} 1 & 0 & 0 & t_x \\ 0 & 1 & 0 & t_y \\ 0 & 0 & 1 & t_z \\ 0 & 0 & 0 & 1 \end{bmatrix}.$$

Problem 5.9 Show that $\mathbf{T}(-t_x, -t_y, -t_z)$ is the inverse of $\mathbf{T}(t_x, t_y, t_z)$. $\mathbf{T}^{-1}$ undoes the work of $\mathbf{T}$. ∎

Rotation may be done about any arbitrary line in three dimensions. We will build up to the general case by first presenting the operators that rotate about the three coordinate axes. $\mathbf{R}_x(\theta)$ rotates by angle θ about the x-axis, with positive θ going from the y-axis to the z-axis, as shown in Figure 5.5. In a similar fashion, positive rotation about the y-axis using $\mathbf{R}_y(\theta)$ goes from z to x, and positive rotation about the z-axis goes from x to y, just as in two dimensions. We have the fundamental rotations

$$\mathbf{R}_x(\theta) = \begin{bmatrix} 1 & 0 & 0 & 0 \\ 0 & \cos\theta & -\sin\theta & 0 \\ 0 & \sin\theta & \cos\theta & 0 \\ 0 & 0 & 0 & 1 \end{bmatrix}$$

$$\mathbf{R}_y(\theta) = \begin{bmatrix} \cos\theta & 0 & \sin\theta & 0 \\ 0 & 1 & 0 & 0 \\ -\sin\theta & 0 & \cos\theta & 0 \\ 0 & 0 & 0 & 1 \end{bmatrix} \qquad 5.24$$

$$\mathbf{R}_z(\theta) = \begin{bmatrix} \cos\theta & -\sin\theta & 0 & 0 \\ \sin\theta & \cos\theta & 0 & 0 \\ 0 & 0 & 1 & 0 \\ 0 & 0 & 0 & 1 \end{bmatrix}.$$

A more general rotation about any line through the origin can be constructed by composition of the three fundamental rotations. Finally, by composing translation with the fundamental rotations, we can build an operator that rotates about any arbitrary line in space.

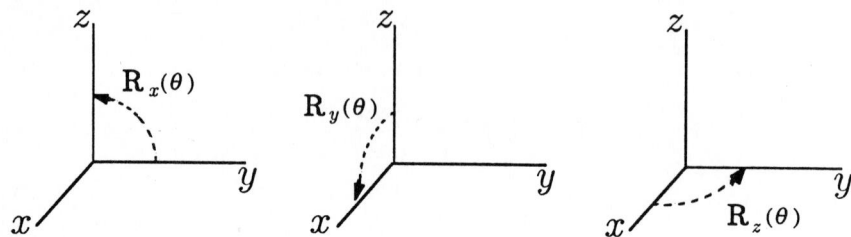

Figure 5.5: Directions of Positive Rotation

Example 5.9. To rotate by angle ϕ about the line $\mathcal{L}$, which lies in the x-y plane in Figure 5.6, we would

(i) rotate $\mathcal{L}$ to the x-axis with $\mathbf{R}_z(-\theta)$;

(ii) rotate by ϕ about the x-axis with $\mathbf{R}_x(\phi)$; and

(iii) rotate back to $\mathcal{L}$ with $\mathbf{R}_z(\theta)$.

The composite operation would be

$$\mathbf{A}(\theta, \phi) = \mathbf{R}_z(\theta)\mathbf{R}_x(\phi)\mathbf{R}_z(-\theta). \quad \square$$

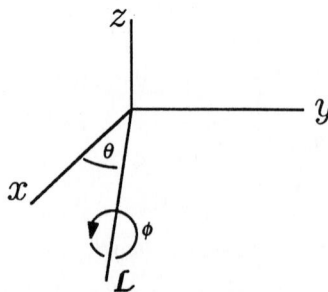

Figure 5.6: Composition of Rotations

Direction Cosines. As discussed in Chapter 4, a vector $\mathbf{v}$ may be specified by its coordinates (x, y, z) or by its length and direction. The length of $\mathbf{v}$ is $\|\mathbf{v}\|$, and the direction can be specified in terms of the three *direction cosines* $(\cos\theta_x, \cos\theta_y, \cos\theta_z)$. The angle θ_x is measured between the vector $\mathbf{v}$ and the x-axis or, equivalently, between the vector $\mathbf{v}$ and the vector $\mathbf{e}_x = [1\ 0\ 0]^T$. We have

$$\cos\theta_x = \frac{(\mathbf{v}, \mathbf{e}_x)}{\|\mathbf{v}\|\,\|\mathbf{e}_x\|} = \frac{x}{\|\mathbf{v}\|}. \qquad 5.25$$

Likewise, θ_y is measured between $\mathbf{v}$ and $\mathbf{e}_y = [0\ 1\ 0]^T$, and θ_z is measured between $\mathbf{v}$ and $\mathbf{e}_z = [0\ 0\ 1]^T$. Thus

$$\cos\theta_y = \frac{y}{\|\mathbf{v}\|} \qquad 5.26$$

$$\cos\theta_z = \frac{z}{\|\mathbf{v}\|}.$$

The vector

$$\mathbf{u} = \begin{bmatrix} \cos\theta_x \\ \cos\theta_y \\ \cos\theta_z \end{bmatrix} \qquad 5.27$$

is a unit vector in the direction of $\mathbf{v}$, so we have

$$\mathbf{v} = \|\mathbf{v}\|\,\mathbf{u} = \|\mathbf{v}\| \begin{bmatrix} \cos\theta_x \\ \cos\theta_y \\ \cos\theta_z \end{bmatrix}. \qquad 5.28$$

Problem 5.10 Show that $\mathbf{u}$ is a unit vector (i.e. $\|\mathbf{u}\| = 1$). ∎

The direction cosines are useful for specifying a line in space. Instead of giving two points P_1 and P_2 on the line, we can give one point P_1 plus the direction cosines of any vector that points along the line. One such vector is the vector from P_1 to P_2.

Arc tangents. Consider a vector $\mathbf{v} = \begin{bmatrix} x \\ y \end{bmatrix}$ in two dimensions. We know that

$$\tan \theta = \frac{y}{x} \qquad\qquad 5.29$$

for the angle θ shown in Figure 5.7. If we know x and y, we can find θ using the arc tangent function

$$\theta = \tan^{-1}\left(\frac{y}{x}\right). \qquad\qquad 5.30$$

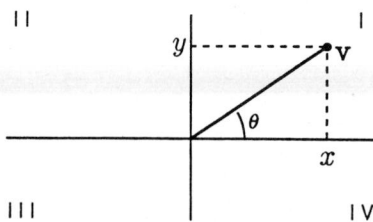

Figure 5.7: Tangent and Arc Tangent

In MATLAB,

theta = atan(y/x)

Unfortunately, the arc tangent always gives answers between $\frac{-\pi}{2}$ and $\frac{\pi}{2}$, corresponding to points $\mathbf{v}$ in quadrants I and IV. The problem is that the ratio $\frac{y}{x}$ is the same as the ratio $\frac{-y}{-x}$, so quadrant III cannot be distinguished from quadrant I by the ratio $\frac{y}{x}$. Likewise, quadrants II and IV are indistinguishable.

The solution is the two-argument arc tangent function. In MATLAB,

theta = atan2(y, x)

will give the true angle θ between $-\pi$ and π in any of the four quadrants.

Example 5.10. Consider the direction vector $\mathbf{u} = \begin{bmatrix} \cos \theta_x \\ \cos \theta_y \\ \cos \theta_z \end{bmatrix}$, as shown in Figure 5.8. What is the angle ϕ_y between the projection of $\mathbf{u}$ into the x-y

plane and the y-axis? This is important because it is $\mathbf{R}_z(\phi_y)$ that will put $\mathbf{u}$ in the y-z plane, and we need to know the angle ϕ_y in order to carry out this rotation. First note that it is *not* the same as θ_y. From the geometry of the figure, we can write

$$\tan \phi_y = \frac{\cos \theta_x}{\cos \theta_y}. \qquad\qquad 5.31$$

This gives us a formula for ϕ_y in terms of the direction cosines of $\mathbf{u}$. With the two-argument arc tangent, we can write

$$\phi_y = \tan^{-1}(\cos \theta_x, \cos \theta_y). \ \Box \qquad\qquad 5.32$$

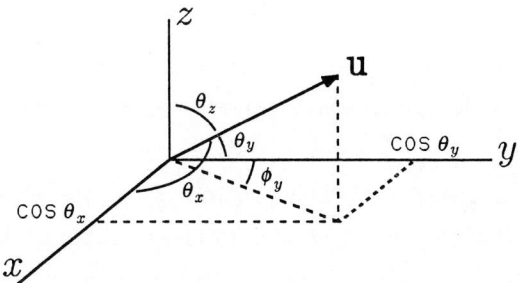

Figure 5.8: Angles in Three Dimensions

Problem 5.11

(a) Suppose point $\mathbf{p}'$ is in the y-z plane in three dimensions, $\mathbf{p}' = (0, y', z', 1)$. Find θ so that $\mathbf{R}_x(\theta)$ will rotate $\mathbf{p}'$ to the positive z-axis. (Hint: Use the two-argument arc tangent. θ will be a function of y' and z'.)

(b) Let $\mathbf{p}$ be any point in three-dimensional space, $\mathbf{p} = (x, y, z, 1)$. Find ψ so that $\mathbf{R}_z(\psi)$ rotates $\mathbf{p}$ into the y-z plane. (Hint: Sketch the situation in three-dimensions, then sketch a two-dimensional view looking down at the x-y plane from the positive z-axis. Compare with Example 5.10.)

Your answers to parts (a) and (b) can be composed into an operator $\mathbf{Z}(\mathbf{p})$ that rotates a given point $\mathbf{p}$ to the positive z-axis, $\mathbf{Z}(\mathbf{p}) = \mathbf{R}_x(\theta)\mathbf{R}_z(\psi)$.

(c) Let $\mathcal{L}$ be a line in three-dimensional space specified by a point $\mathbf{l} = (x, y, z, 1)$ and the direction cosines $(\cos\theta_x, \cos\theta_y, \cos\theta_z)$. Use the following procedure to derive a composite operator $\mathbf{R}(\phi, \mathcal{L})$ that rotates by angle ϕ about the line $\mathcal{L}$:

(i) translate $\mathbf{l}$ to the origin;

(ii) let $\mathbf{u} = (\cos\theta_x, \cos\theta_y, \cos\theta_z, 1)$ and use $\mathbf{Z}(\mathbf{u})$ to align $\mathcal{L}$ with the z-axis;

(iii) rotate by ϕ about the z-axis;

(iv) undo step (ii); and

(v) undo step (i). ∎

5.7 Projections

Computer screens and printers are two-dimensional display devices. We must somehow convert three-dimensional images to two dimensions in order to display them. This task is done by another kind of matrix operator called a *projection*.

To build a projection, we first choose a *projection plane* in the three-dimensional space of the object we wish to view. All points of the object are then projected onto the plane. There are many different kinds of projections, corresponding to various geometric rules for mapping points in space onto a plane. We begin with the *parallel projection* illustrated in Figure 5.9, wherein the dotted lines between the points and their projections in the plane are all parallel to one another. These dotted lines are called *projectors*.

Range. The projection plane is called the *range* of the projection. We will assume for now that the projection plane passes through the origin. If it does not, we may later compose the necessary translations with our projection. Three points, not all in a line, are required to determine a plane. We will take the origin as one of the points and suppose that the other two points are the vectors $\mathbf{r}_1$ and $\mathbf{r}_2$. From these vectors, we form the matrix

$$\mathbf{R} = [\,\mathbf{r}_1\ \mathbf{r}_2\,], \qquad\qquad 5.33$$

which determines the range of the projection.

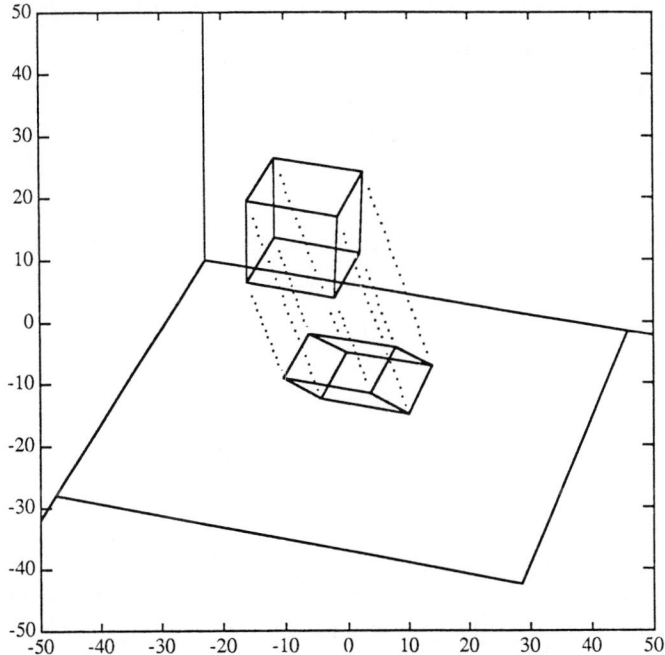

Figure 5.9: Oblique Parallel Projection

Null-Space. To complete the specification of the parallel projection, we must specify the direction of the projectors. The line through the origin in this direction is called the *null-space* of the projection. Note that any line in the three-dimensional object that is parallel to the null-space will disappear to a point when projected. The null-space may be specified by a vector **n** pointing in the direction of the projectors.

Orthogonal Projection. If the null-space is perpendicular to the range, we say that the projection is an *orthogonal projection*. The matrix operator for orthogonal projection is a function of the range. If we assume that **R** is specified in three-dimensional coordinates (not homogeneous!), we have the following definition for the orthogonal projection onto the range of **R**:

$$\mathbf{P}(\mathbf{R}) = \mathbf{R}(\mathbf{R}^T\mathbf{R})^{-1}\mathbf{R}^T \ \in \ \mathcal{R}^{3\times3}. \qquad 5.34$$

In homogeneous coordinates, we embed this 3×3 projection matrix in the general 4×4 transformation:

$$\mathbf{P}_h(\mathbf{R}) = \begin{bmatrix} \mathbf{P}(\mathbf{R}) & \mathbf{0} \\ \mathbf{0}^T & 1 \end{bmatrix}.$$ (5.35)

Example 5.11. Suppose we choose the x-y plane as our projection plane. Vectors $\mathbf{r}_1$ and $\mathbf{r}_2$ can be any two points in the plane (besides the origin), so let's take

$$\mathbf{r}_1 = \begin{bmatrix} 1 \\ 0 \\ 0 \end{bmatrix}, \qquad \mathbf{r}_2 = \begin{bmatrix} 1 \\ 1 \\ 0 \end{bmatrix}.$$

Then the range matrix is

$$\mathbf{R} = \begin{bmatrix} 1 & 1 \\ 0 & 1 \\ 0 & 0 \end{bmatrix}.$$

The orthogonal projection looking down on the x-y plane from the z-axis is

$$\mathbf{P}(\mathbf{R}) = \begin{bmatrix} 1 & 1 \\ 0 & 1 \\ 0 & 0 \end{bmatrix} \left(\begin{bmatrix} 1 & 0 & 0 \\ 1 & 1 & 0 \end{bmatrix} \begin{bmatrix} 1 & 1 \\ 0 & 1 \\ 0 & 0 \end{bmatrix} \right)^{-1} \begin{bmatrix} 1 & 0 & 0 \\ 1 & 1 & 0 \end{bmatrix}$$

$$= \begin{bmatrix} 1 & 0 & 0 \\ 0 & 1 & 0 \\ 0 & 0 & 0 \end{bmatrix}.$$

In homogeneous coordinates,

$$\mathbf{P}_h(\mathbf{R}) = \begin{bmatrix} 1 & 0 & 0 & 0 \\ 0 & 1 & 0 & 0 \\ 0 & 0 & 0 & 0 \\ 0 & 0 & 0 & 1 \end{bmatrix}. \quad \square$$

Problem 5.12 Let the vectors $\mathbf{r}_1 = \begin{bmatrix} 2 \\ 0 \\ 0 \end{bmatrix}$ and $\mathbf{r}_2 = \begin{bmatrix} 0 \\ 3 \\ 0 \end{bmatrix}$ specify the range of an orthogonal projection. Find $\mathbf{P}(\mathbf{R})$ and $\mathbf{P}_h(\mathbf{R})$. Compare with Example 5.11 and explain any similarities. ∎

Example 5.12. In Chapter 4 you learned that the projection of $\mathbf{w}$ onto $\mathbf{x}$ is given by

$$\mathbf{z} = \frac{(\mathbf{x}, \mathbf{w})\mathbf{x}}{(\mathbf{x}, \mathbf{x})}.$$

This is an orthogonal projection of $\mathbf{w}$ onto a line, but it is closely related to the projection into the plane just described. To see the similarity, let's work on the expression for $\mathbf{z}$ a little:

$$\begin{aligned}
\mathbf{z} &= \frac{\mathbf{x}(\mathbf{x}, \mathbf{w})}{(\mathbf{x}, \mathbf{x})} \\
&= \mathbf{x}(\mathbf{x}, \mathbf{x})^{-1}(\mathbf{x}, \mathbf{w}) \\
&= \mathbf{x}(\mathbf{x}^T\mathbf{x})^{-1}(\mathbf{x}^T\mathbf{w}) \\
&= \left[\mathbf{x}(\mathbf{x}^T\mathbf{x})^{-1}\mathbf{x}^T\right]\mathbf{w} \\
&= \mathbf{P}(\mathbf{x})\mathbf{w}.
\end{aligned}$$

So we see that it can be written in the same way as the projection onto the plane. The only difference is that the range is now one-dimensional and specified by $\mathbf{x}$ in place of $\mathbf{R}$. □

Oblique Projection. More generally, we may have a null-space $\mathbf{n}$ that is not perpendicular to the range $\mathbf{R}$. The projection shown in Figure 5.9 is an oblique projection. Once again we start with nonhomogeneous coordinates for $\mathbf{n}$ and $\mathbf{R}$ and write the 3×3 oblique projection as

$$\mathbf{E}(\mathbf{R}, \mathbf{n}) = \mathbf{R}\{\mathbf{R}^T[\mathbf{I} - \mathbf{P}(\mathbf{n})]\mathbf{R}\}^{-1}\mathbf{R}^T[\mathbf{I} - \mathbf{P}(\mathbf{n})] \qquad 5.36$$

where

$$\mathbf{P}(\mathbf{n}) = \mathbf{n}(\mathbf{n}^T\mathbf{n})^{-1}\mathbf{n}^T = \frac{\mathbf{n}\mathbf{n}^T}{\|\mathbf{n}\|^2}.$$

As with the orthogonal projection, we can return to homogeneous coordinates by

$$\mathbf{E}_h(\mathbf{R}, \mathbf{n}) = \begin{bmatrix} \mathbf{E}(\mathbf{R}, \mathbf{n}) & \mathbf{0} \\ \mathbf{0}^T & 1 \end{bmatrix}.$$
5.37

Problem 5.13 Prove and interpret the following properties of parallel projections (both orthogonal and oblique):

(a) $\mathbf{P}^2 = \mathbf{P}$; $\mathbf{E}^2 = \mathbf{E}$.

(b) $\mathbf{PR} = \mathbf{R}$; $\mathbf{ER} = \mathbf{R}$.

(c) $\mathbf{En} = \mathbf{0}$; $\mathbf{Pn} = \mathbf{0}$. (First show that $\mathbf{R}^T\mathbf{n} = \mathbf{0}$ when $\mathbf{n}$ is orthogonal to $\mathbf{R}$.) ∎

Display from Projections. Even after we have used a projection, the image points are in three-dimensional homogeneous coordinates. How then do we display them on a two-dimensional display? If we had chosen the x-y plane as the range of our projection, we could let the output device represent the x-y plane and ignore the z-coordinate of each point. The z-coordinates of the projected image would all be 0 anyway since the projected points would lie in the x-y plane. The fourth coordinate of 1 may also be ignored for display. But even if the projection plane is not the x-y plane, we may perform a rotation to align the two planes and again let the output device represent the x-y plane.

Perspective Projections. To obtain a perspective projection, we first choose a projection plane as we did for parallel projections. Instead of choosing a null-space parallel to all projectors, we now choose a viewpoint. The projectors all pass through the viewpoint in a perspective projection, as shown in Figure 5.10. For a viewpoint at $z = d$ on the z-axis and a projection plane coinciding with the x-y plane, the three-dimensional homogeneous operator for perspective projection is

$$\mathbf{M}(d) = \begin{bmatrix} 1 & 0 & 0 & 0 \\ 0 & 1 & 0 & 0 \\ 0 & 0 & 0 & 0 \\ 0 & 0 & \frac{-1}{d} & 1 \end{bmatrix}.$$
5.38

The first thing you should notice about the perspective projection $\mathbf{M}(d)$ is that it violates the structure given in Equation 5.22 by having the 4,3 position equal to $-\frac{1}{d}$ rather than to 0. This means that the fourth coordinates will not remain 1 in the new point matrix

$$\mathbf{G}_{\text{new}} = \mathbf{M}(d)\mathbf{G}.$$

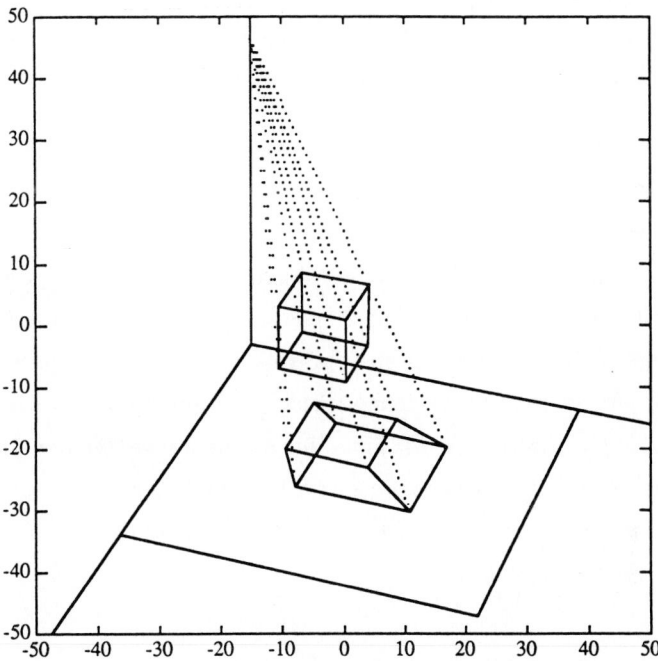

Figure 5.10: A Perspective Projection

This situation is interpreted to mean that the new point (x, y, z, w) must be *renormalized* to $\left(\frac{x}{w}, \frac{y}{w}, \frac{z}{w}, 1\right)$ before the operation is considered complete. Renormalization adds another computational step to the image transformation process. This is undesirable, but we are stuck with it if we wish to include perspective projections in our repertoire of transformations. Note that renormalization is a point-by-point process since w may be different for each point.

How does renormalization affect composition of operators? We might expect that, each time we perform an operation requiring renormalization, we will be forced to stop and do the renormalization before going on with other operations. In this respect we are fortunate: we may put off the renormalization until all transformations are complete and renormalize once just before displaying the image. Thus we are able to compose perspective projections at will with other transformations.

Problem 5.14 What is the perspective projection matrix for a viewpoint at infinity, $\mathbf{M}(\infty)$? Interpret the result. ∎

Generalization of Projections. "Projection" is a technical term in linear algebra. A square matrix of any size is considered a projection if it obeys the property of Problem 5.13(a), $\mathbf{P}^2 = \mathbf{P}$. All of our geometric projection matrices have this property. We have considered only three-dimensional spaces with a two-dimensional range and a one-dimensional null-space. In general, the dimensionality of the space may be split in any way between range and null-space. If we reinterpret homogeneous coordinates as four-dimensional space, we have projections with three-dimensional range and one-dimensional null-space, but the perspective projection is actually not quite a projection in the technical sense even though $\mathbf{M}^2(d) = \mathbf{M}(d)$ because perspective projection also includes renormalization.

5.8 Numerical Experiment (Star Field)

With Earth at the origin, we may specify the star positions for the Big and Little Dippers in three-dimensional homogeneous coordinates. With light years as our units, we have $\mathbf{G}^T =$

1.5441	-1.2064	153.0875	1.0000
-1.0386	8.4588	142.7458	1.0000
-8.7955	26.2870	198.0698	1.0000
-12.8807	19.0964	106.5383	1.0000

-18.8926	17.461	90.6185	1.0000
-45.1364	54.2706	215.1148	1.0000
-9.6222	20.1734	88.0062	1.0000
-33.7097	-8.4048	64.6574	1.0000
-33.7144	-8.7191	52.3806	1.0000
-43.8531	-1.1483	59.7193	1.0000
-36.1086	2.3667	55.7927	1.0000
-34.2309	8.2181	52.1260	1.0000
-30.7876	11.8182	46.9810	1.0000
-61.8581	31.5183	80.7615	1.0000

To make use of this data, we need a function to display it. Enter and save the following generalization of the function from Demo 5.1. Call it vhgraph.m.

```
function vhgraph(P,L,A,ps,ls);
% vhgraph(P,L,A,PS,LS) graphs images whose points are
% stored in P and whose lines are stored in L. The points
% in P must be in homogeneous coordinates in either 2 or 3
% dimensions, with each column of P representing a point.
% The lines are coded in L, with each column of L containing
% 2 integers pointing to a pair of points in P to be
% connected by a line segment.  If A is present, the points
% in P are transformed to A*P before graphing.  For 3D data
% points, only the first two coordinates are graphed, so A
% should include the desired projection from 3D to 2D. The
% point symbol may be specified in PS and the line type in
% LS, if desired.
%
% Richard T. Behrens, October 1989.
%
% The first section of the program determines the sizes of
% all the input matrices and checks that they make sense.
[mp,np] = size(P);
if (nargin > 1)
   [ml,nl] = size(L);
else
   ml = 2; nl = 0;
end
```

```
if (nargin <= 2)
   A = eye(mp);
end
[mA,nA] = size(A);
if ((mp ~= nA) | (ml ~= 2))
   error('Incompatible sizes of input matrices.")
end
if (nargin <= 3);
   ps = '*';
end
if (nargin <= 4);
   ls = '-';
end

P = A*P;      % Performs the transformation A on the points
              % (effect is only local to this function).

% The next section contains a loop that renormalizes the
% homogeneous coordinates of the points if necessary.
renorm = find((P(mA,:) ~= 1));
if ~isempty(renorm)
   for i = 1:length(renorm)
      P(:,renorm(i)) = P(:,renorm(i))/P(mA,renorm(i));
   end
end

% The next program line sets a fixed scale output window
% from -50 to 50 in both x and y directions on the screen.
% For automatic scaling to include all points of the
% image, we could use instead the line q = min(min(P));
% r = max(max(P));
q = -50; r = 50;
axis([q r q r])
axis('square')
plot(P(1,:),P(2,:),ps)  % Plots the points with symbol ps.
hold on                 % Saves the points while we plot
for i = 1:nl            % lines with line type LS.
plot([P(1,L(1,i))P(1,L(2,i))], ..
     [P(2,L(1,i))P(2,L(2,i))],ls)
end
hold off
```

Enter the point matrix given at the beginning of this section (and take its transpose to put it in the usual form). Also enter the line matrix from Demo 5.1. Save these two matrices and try looking at the image

```
≫ save dip3d
≫ vhgraph(G,H)
```

No dippers in sight, right? Without specifying a transformation matrix $\mathbf{A}$, we have defaulted to looking down on the x-y plane from $z = \infty$ (a parallel projection). This is how the constellations would look from a distant galaxy (say, a billion light years north of here) through an enormous telescope. We need a perspective view from the origin (Earth), but first we need a set of functions to give us the fundamental operators with which we can build the desired projection.

Take $\mathbf{R}_y(\theta)$ as an example. The function to build it looks like

```
function Ry = vhry(theta);
% Rotation matrix about the y-axis for 3-D homogeneous
% coordinates.
Ry = eye(4);
Ry(1,1) = cos(theta);
Ry(3,3) = cos(theta);
Ry(3,1) = -sin(theta);
Ry(1,3) = sin(theta);
```

Enter and save vhry.m as given. Then write functions for

$\mathbf{R}_x(\theta)$	vhrx.m
$\mathbf{R}_z(\theta)$	vhrz.m
$\mathbf{S}(s_x, s_y, s_z)$	vhs.m
$\mathbf{T}(t_x, t_y, t_z)$	vht.m
$\mathbf{M}(d)$	vhm.m

Useful MATLAB functions for this task include **zeros**, **eye**, and **diag**.

Now build and use a perspective projection with viewpoint Earth and projection plane at $z = 1000$ behind the dippers:

(1) translate Earth to $z = -1000$ so that the projection plane coincides with the x-y plane: $\mathbf{T}(0, 0, -1000)$;

(2) use the fundamental perspective projection: $\mathbf{M}(-1000)$; and

(3) translate back: $\mathbf{T}(0,0,1000)$.

```
>> A = vht(0,0,1000) * vhm(-1000) * vht(0,0,-1000)
>> vhgraph(G,H,A)
```

Oops! Now the image is too big; it's mostly off the screen. Scale it down and have another look:

```
>> A = vhs(.06,.06,.06) * A
>> vhgraph(G,H,A)
```

Now the view should look familiar. Leave $\mathbf{A}$ as it is now:

```
>> save dip3d
```

Experiment with scale and rotation about the z-axis. For example, try

```
>> vhgraph(G,H,vhrz(pi/2) * A)
```

The two-dimensional star positions given in Demo 5.1 were obtained from the three-dimensional positions with the composite operator $\mathbf{A}$ you are now using. To compare the two, type

```
>> Gnew = A * G

>> for i = 1:14
Gnew(:,i) = Gnew(:,i)/Gnew(4,i);
end

>> Gnew
```

and compare the x and y coordinates with those of Demo 5.1.

Astronomers give star positions in equatorial coordinates using right ascension, declination, and distance. The following function converts equatorial coordinates, which are spherical, to Cartesian coordinates with the z-axis pointing north, the x-axis pointing at the vernal (Spring) equinox in the constellation Pisces, and the y-axis pointing toward the Winter solstice in the constellation Opheuchus.

```
function v = starxyz(rah,ram,decd,decm,dist)
% starxyz is the cartesian coordinates of a star whose
% spherical coordinates (e.g. from a star catalog) are
%
%           rah     right ascension hours
%           ram     right ascension minutes
%           decd    declination degrees
%           decm    declination minutes (should be negative
%                   if decd is negative)
%           dist    distance (light years)
%
phi = (pi/180) * (decd + decm/60);
theta = (pi/12) * (rah + ram/60);
r = dist;
v = [r * cos(phi) * cos(theta); -r * cos(phi) * sin(theta);
    r * sin(phi)];
```

Have you ever wondered what the constellations would look like from other places in the galaxy? We will soon see the answer. First we will view the dippers from Alpha Centauri, the nearest star, whose coordinates are

$$-1.5680 \qquad 1.3157 \qquad -3.6675.$$

We will look toward the *centroid* of the fourteen stars in the dippers, located at

$$-26.3632 \qquad 12.8709 \qquad 100.4714.$$

To get the desired view, we must

(1) translate the viewpoint to the origin;

(2) rotate the centroid (direction to look) to the z-axis—note that the centroid will have new coordinates after step (1); and

(3) apply the composite A=S(.06,.06,.06)*T(0,0,1000)*M(-1000)*T(0,0,-1000) (as used to view from Earth).

Write a function vhz.m based on Problem 5.11 to accomplish step (2). Test it on several random points to make sure it works right. Now write a general perspective projection function called vhview.m. The function vhview should accept as inputs two vectors, the first specifying the viewpoint and the second the point to look toward. Its output should be a composite operator that performs all three of the preceding steps.

Now we want to look toward the centroid of the dippers from Alpha Centauri. To do so, enter the vectors for the two points of interest and construct the view like this:

```
≫ acentauri = [-1.5680; 1.3157; -3.6675]
≫ centroid = [-26.3632; 12.8709; 100.4714]
≫ A = vhview(acentauri,centroid)
≫ vhgraph(G,H,A)
≫ title('From Alpha Centauri')
```

The farther we move from Earth, the more distorted the dippers will look in general. It should be easy now to view them from any desired location. Just choose a viewpoint, recalculate the composite operator for that viewpoint using vhview, and use vhgraph. Follow this procedure to view the dippers from each of the stars in the following list. You will need to use starxyz first to convert their coordinates.

Star	Right Ascension	Declination	Distance (ly)
Alpha Centauri	14h 40m	−60° 50′	4.2
Sirius	6h 45m	−16° 43′	9.5
Arcturus	14h 16m	19° 11′	16.6
Pollux	7h 45m	28° 02′	35.9
Betelgeuse	5h 55m	7° 24′	313.5

Of course, star viewing is not the only application of vector graphics. Do some experiments with the unit cube (see Problem 5.8). View the cube from location (4,3,2) looking toward the origin using the procedure just outlined for stars. You may need to adjust the scaling to get a meaningful view.

6

| | | | | | | | | | | | **6** |

Filtering

Notes to Teachers and Students:

Filtering is one of the most important things that electrical and computer engineers do. In this chapter we extend everyday understanding of filters to *numerical* filters. We then study weighted moving averages and exponential averages. We define the important *test signals* for electrical and computer engineering and show how filters respond to them. The idea that filters are characterized by their response to simple test signals is fundamental. In the numerical experiment in Section 6.7, students explore the *frequency response* of a simple filter, a concept that forms the basis of circuit theory, electronics, optics and lasers, solid-state devices, communications, and control.

6.1 Introduction

A *filter* is any device that passes material, light, sound, current, velocity, or information according to some rule of selectivity. Material (or mechanical) filters are commonplace in your everyday life:

(i) coffee filters pass flavored water while filtering out coffee grounds;

(ii) Goretex fibers pass small, warm perspiration droplets while filtering out large, cool droplets of rain or snow;

(iii) fiberglass strands in a furnace filter pass warm air while filtering out particles of dirt and dust;

(iv) a centrifuge retains material of low density while spinning out (or filtering out) material of high density; and

(v) an electrostatic precipitator filters out dust and other effluents by attaching charge to them and using an electric field to move the charged particles to a high potential drain.

The first three of these examples selectively pass material according to size; the last two selectively pass material according to its mass density.

Typical filters for light are

(i) UV filters on camera lenses and eyeglasses that pass light in the range of visible wavelengths while blocking light in the invisible (but damaging) ultraviolet range;

(ii) polaroid lenses that pass light that is randomly polarized while blocking
out glare that is linearly polarized;

(iii) green fabrics that reflect green light and absorb other colors;

(iv) red taillights that pass light in the long wavelength red range and reflect
light in the short wavelength violet range (look at the inside of your
taillights to see violet); and

(v) glacial ice that absorbs all but the blue wavelengths so that it appears
blue.

Problem 6.1 List as many examples of natural and man-made sound filters
as you can. ∎

Satellite Television. Among current filters, the tuner in a superheterodyne receiver is, perhaps, the first example that comes to mind. But
satellite TV filters are another fascinating example. A typical C-band satellite has twelve transponders (or repeaters), each of which transmits microwave
radiation in a personalized 36 MHz band. (The abbreviation MHz stands for
megahertz, or 10^6 Hz, or 10^6 cycles per second. Other common abbreviations
are Hz for 1 Hz, kHz for 10^3 Hz, and GHz for 10^9 Hz.) Each transponder actually transmits two channels of information, one vertically polarized and one
horizontally polarized. There is an 8 MHz guard band between each band,
and the vertical and horizontal channels are offset by 20 MHz. The transmission scheme for the 24 channels is illustrated in Figure 6.1. The entire
transmission band extends over 540 MHz, from 3.7×10^9 Hz to 4.24×10^9 Hz.
The satellite receiver has two different microwave detectors, one for vertical
and one for horizontal polarization, and a microwave tuner to tune into the
microwave band of interest.

Problem 6.2 Check that the transmission scheme of Figure 6.1 consumes
540 MHz of bandwidth. ∎

Problem 6.3 List as many examples of natural and man-made devices for
velocity filtering as you can. ∎

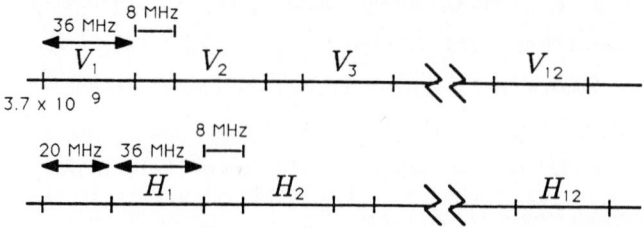

Figure 6.1: Satellite TV— V_i, Vertically Polarized Channel i; H_j, Horizontally Polarized Channel j

An Aside on Hertz and Seconds. The abbreviation Hz stands for hertz, or cycles/second. It is used to describe the frequency of a sinusoidal signal. For example, house current is 60 Hz, meaning that it has 60 cycles each second. The inverse of Hz is seconds or, more precisely, seconds/cycle, the period of 1 cycle. For example, the period of 1 cycle for house current is 1/60 second. When we are dealing with sound, electricity, and electromagnetic radiation, we need a concise language for dealing with signals and waves whose frequencies range from 0 Hz (called DC or direct current) to 10^{18} Hz (visible light). Table 6.1 summarizes the terms and symbols used to describe the frequency and period of signals that range in frequency from 0 Hz to 10^{12} Hz.

Table 6.1: Terms and Symbols for Sinusoidal Signals

Frequency			Period			
Hz	Term	Units	Seconds	Term	Units	Example
Hz	hertz	1 Hz	sec	second	1 sec	battery current: 0 Hz
						house current: 60 Hz
kHz	kilohertz	10^3 Hz	msec	millisecond	10^{-3} sec	midfrequency sound
MHz	megahertz	10^6 Hz	μsec	microsecond	10^{-6} sec	clock frequencies in
						microcomputers
GHz	gigahertz	10^9 Hz	nsec	nanosecond	10^{-9} sec	microwave radiation for
						satellite communication
THz	terahertz	10^{12} Hz	psec	picosecond	10^{-12} sec	infrared radiation

Numerical Filters. Rather amazingly, these ideas extend to the domain of numerical filters, the topic of this chapter. Numerical filters are just schemes for weighting and summing strings of numbers. Stock prices are typically averaged with numerical filters. The curves in Figure 6.2 illustrate the daily closing average for Kellogg's common stock and two moving averages. The 50-day moving average is obtained by passing the daily closing average through a numerical filter that averages the most current 50 days' worth of closing averages. The 200-day moving average for the stock price is obtained by passing the daily closing prices through a numerical filter that averages the most current 200 days' worth of daily closing averages. The daily closing averages show fine-grained variation but tend to conceal trends. The 50-day and 200-day averages show less fine-grained variation but give a clearer picture of trends. In fact, this is one of the key ideas in numerical filtering: by selecting our method of averaging, we can filter out fine-grained variations and pass long-term trends (or vice versa), or we can filter out periodic variations and pass nonperiodic variations (or vice versa). Figure 6.2 illustrates that moving averages typically lag increasing sequences of numbers and lead decreasing sequences. Can you explain why?

We will call any algorithm or procedure for transforming one set of numbers into another set of numbers a *numerical filter* or *digital filter*. Digital filters, consisting of memories and arithmetic logic units (ALUs), are implemented in VLSI circuits and used for communication, control, and instrumentation. They are also implemented in random—or semicustom—logic circuits and in programmable microcomputer systems. The inputs to a digital filter are typically electronic measurements that are produced by A/D (analog-to-digital) conversion of the output of an electrical or mechanical sensor. The outputs of the filter are "processed," "filtered," or "smoothed" versions of the measurements. In your more advanced courses in electrical and computer engineering you will study signal processing and system theory, assembly language programming, microprocessor system development, and computer design. In these courses you will study the design and programming of hardware that may be used for digital filtering.

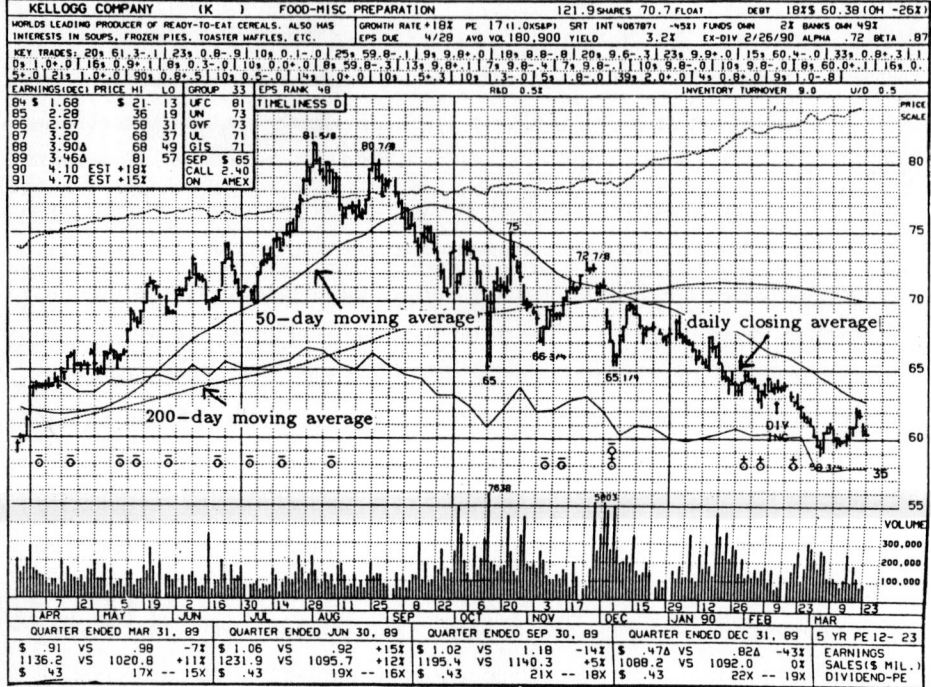

Figure 6.2: Dow-Jones Averages (Adapted from the New York Stock Exchange, *Daily Graphs*, William O'Neil and Co., Inc., Los Angeles, California)

6.2 Simple Averages

The simplest numerical filter is the simple averaging filter. This filter is defined by the equation

$$x = \frac{1}{N} \sum_{n=1}^{N} u_n. \qquad 6.1$$

The filter output x is the average of the N filter inputs $u_1, u_2, \ldots, u_N$. These inputs may be real or complex numbers, and x may be real or complex. This simple averaging filter is illustrated in Figure 6.3.

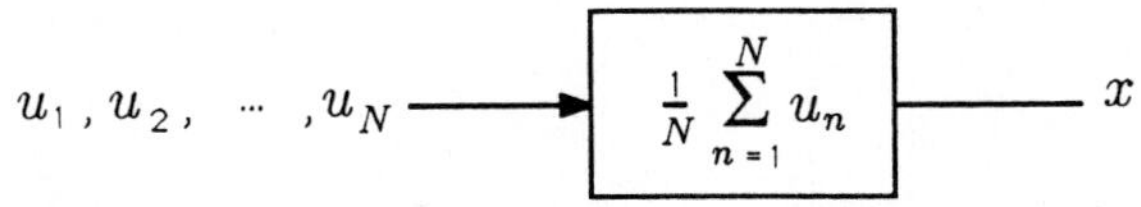

$$u_1, u_2, \cdots, u_N \longrightarrow \boxed{\frac{1}{N} \sum_{n=1}^{N} u_n} \longrightarrow x$$

Figure 6.3: A Simple Averaging Filter

Example 6.1. If the averaging filter is excited by the constant sequence $u_1 = u_2 = \cdots = u_N = u$, then the output is

$$x = \frac{1}{N} \sum_{n=1}^{N} u = u. \qquad 6.2$$

The output is, truly, the average of the inputs. Now suppose the filter is excited by the linearly increasing sequence

$$u_n = n, \qquad n = 1, 2, \ldots, N. \qquad 6.3$$

This sequence is plotted in Figure 6.4. How do we sum such a sequence in order to produce the average x? For N even, the average may be written as

$$x = \frac{1}{N}(x_1 + x_N) + \frac{1}{N}(x_2 + x_{N-1}) + \cdots + \frac{1}{N}(x_{N/2} + x_{(N/2)+1}). \qquad 6.4$$

Each pair-sum in parentheses equals $N + 1$, and there are $\frac{N}{2}$ such pair-sums, so the average is

$$x = \frac{1}{N} \frac{N}{2}(N + 1) = \frac{N + 1}{2}. \qquad 6.5$$

This is certainly a reasonable answer for the average of a linearly increasing sequence. See Figure 6.4. □

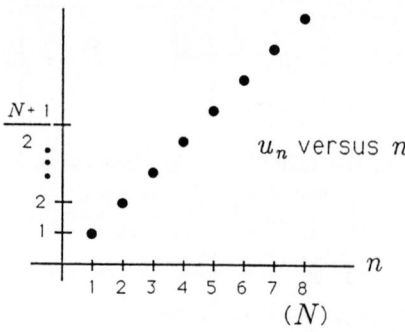

N + 1
2
.
.
.
2
1

u_n versus n

1 2 3 4 5 6 7 8

n

(N)

Figure 6.4: Linearly Increasing Sequence

Problem 6.4   Write $x = \frac{1}{N} \sum_{n=1}^{N} n$ as a sum of pair-sums for N odd. What does x equal? ∎

General Sum Formula. Suppose the input to the simple averaging filter is the polynomial sequence

$$u_n = n^k, \qquad n = 1, 2, \ldots, N \qquad\qquad 6.6$$

where k is a non-negative integer such as $k = 0, 1, 2, \ldots$. The output of the filter is

$$x_N^{(k)} = \frac{1}{N} \sum_{n=1}^{N} n^k. \qquad\qquad 6.7$$

We rewrite x as $x_N^{(k)}$ to remind ourselves that we are averaging N numbers, each of which is n^k. For example, when $N = 8$ and $k = 2$,

$$x_8^{(2)} = \frac{1}{8} \sum_{n=1}^{8} n^2 = \frac{1}{8}(1 + 4 + 9 + \cdots + 64). \qquad\qquad 6.8$$

Rather than study the average $x_N^{(k)}$, we will study the sum $N x_N^{(k)}$ and divide by N at the very end:

$$S_N^{(k)} = N x_N^{(k)} = \sum_{n=1}^{N} n^k. \qquad\qquad 6.9$$

The sum $S_N^{(k)}$ may be rewritten as the sum

$$S_N^{(k)} = \sum_{n=1}^{N-1} n^k + N^k$$

$$= S_{N-1}^{(k)} + N^k.$$

6.10

This result is very important because it tells us that the sum $S_N^{(k)}$, viewed as a function of N, obeys a *recursion* in which $S_N^{(k)}$ is just the sum using one less input, namely, $S_{N-1}^{(k)}$, plus N^k. Now, since polynomials are the most general functions that obey such recursions, we know that $S_N^{(k)}$ must be a polynomial of order $k + 1$ in the variable N:

$$S_N^{(k)} = a_0 + a_1 N + a_2 N^2 + \cdots + a_{k+1} N^{k+1}.$$

6.11

Let's check to see that this polynomial really can obey the required recursion. First note that $S_{N-1}^{(k)}$ is the following polynomial:

$$S_{N-1}^{(k)} = a_0 + a_1(N - 1) + \cdots + a_{k+1}(N - 1)^{k+1}.$$

6.12

The term $(N - 1)^{k+1}$ produces $\binom{k+1}{0} N^{k+1}(-1)^0 + \binom{k+1}{1} N^k(-1)^1 + \cdots$. (Remember the binomial expansion?) Therefore the difference between $S_N^{(k)}$ and $S_{N-1}^{(k)}$ is

$$S_N^{(k)} - S_{N-1}^{(k)} = c_0 + c_1 N + \cdots + c_k N^k.$$

6.13

This recursion is general enough to produce the difference N^k provided we can solve for $a_0, a_1, \ldots, a_{k+1}$ to make $c_0 = c_1 = \cdots = c_{k-1} = 0$ and $c_k = 1$. We know that $S_N^{(k)} = 0$ for $N = 0$, so we know that $a_0 = 0$, meaning that the polynomial for $S_N^{(k)}$ can really be written as

$$S_N^{(k)} = a_1 N + a_2 N^2 + \cdots + a_{k+1} N^{k+1}.$$

6.14

In order to solve for the coefficients of this polynomial, we propose to write out our equation for $S_N^{(k)}$ as follows:

$$(N = 1) \qquad S_1^{(k)} = a_1 + \cdots + a_{k+1}$$

$$(N = 2) \qquad S_2^{(k)} = 2a_1 + \cdots + 2^{k+1}a_{k+1}$$

$$(N = 3) \qquad S_3^{(k)} = 3a_1 + \cdots + 3^{k+1}a_{k+1}$$

$$\vdots \qquad\qquad \vdots \qquad\qquad\qquad\qquad 6.15$$

$$(N = k) \qquad S_k^{(k)} = ka_1 + \cdots + k^{k+1}a_{k+1}$$

$$(N = k+1) \qquad S_{k+1}^{(k)} = (k+1)a_1 + \cdots + (k+1)^{k+1}a_{k+1}.$$

Using the linear algebra we learned in Chapter 4, we may write these equations as the matrix equation

$$\begin{bmatrix} 1 & 1 & \cdots & 1 \\ 2 & 4 & \cdots & 2^{k+1} \\ \vdots & \vdots & & \vdots \\ k & k^2 & \cdots & k^{k+1} \\ (k+1) & (k+1)^2 & \cdots & (k+1)^{k+1} \end{bmatrix} \begin{bmatrix} a_1 \\ a_2 \\ \vdots \\ a_{k+1} \end{bmatrix} = \begin{bmatrix} S_1^{(k)} \\ S_2^{(k)} \\ \vdots \\ S_{k+1}^{(k)} \end{bmatrix}. \qquad 6.16$$

The terms on the right-hand side of the equal sign are "initial conditions" that tell us how the sum $S_N^{(k)}$ begins for $N = 1, 2, \ldots, k+1$. These initial conditions must be computed directly. (For example, $S_2^{(k)} = 1^k + 2^k$.) Then the linear system of $(k+1)$ equations in $(k+1)$ unknowns may be solved for $a_1, a_2, \ldots, a_{k+1}$. The solution for S_N^k is then complete, and we may use it to solve for S_N^k for arbitrary N.

Example 6.2. When $k = 2$, we have the following equation for the coefficients a_1, a_2, and a_3 in the polynomial $S_N^{(2)} = a_1N + a_2N^2 + a_3N^3$:

$$\begin{bmatrix} 1 & 1 & 1 \\ 2 & 4 & 8 \\ 3 & 9 & 27 \end{bmatrix} \begin{bmatrix} a_1 \\ a_2 \\ a_3 \end{bmatrix} = \begin{bmatrix} 1^2 \\ 1^2 + 2^2 \\ 1^2 + 2^2 + 3^2 \end{bmatrix} = \begin{bmatrix} 1 \\ 5 \\ 14 \end{bmatrix}. \;\square \qquad 6.17$$

Problem 6.5 Solve for a_1, a_2, a_3 in the linear equation of Example 6.2. Show that $S_N^{(2)} = a_1N + a_2N^2 + a_3N^3$ obeys the recursion $S_N^{(2)} - S_{N-1}^{(2)} = N^2$. ∎

Problem 6.6 (MATLAB) Write a MATLAB program to determine the coefficients $a_1, a_2, \ldots, a_{k+1}$ for the polynomial $S_N^{(k)}$. Generate a table of formulas for the averages $x_N^{(k)}$ for $k = 1, 2, \ldots, 5$. Evaluate these formulas for $N = 2$, 4, 8, and 16. ∎

Exponential Sums. When the input to an averaging filter is the sequence

$$u_n = a^n, \qquad n = 0, 1, 2, \ldots, N-1,$$ 6.18

we say that the input is exponential (or geometric). Typical sequences are illustrated in Figure 6.5 for $a = 0.9$, $a = 1$, and $a = 1.1$. Don't let it throw you that we have changed the index to run from 0 to $N-1$ rather than from 1 to N. This change is not fundamentally important, but it simplifies our study. The sum of the inputs is

$$S_N = \sum_{n=0}^{N-1} a^n.$$ 6.19

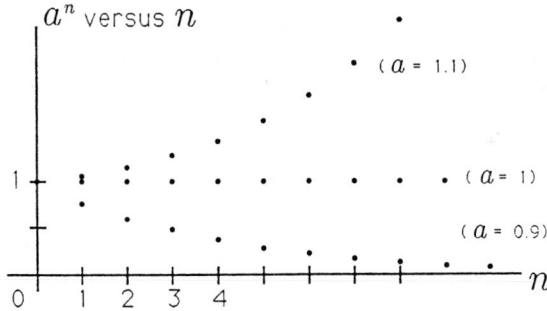

Figure 6.5: Exponential Sequences

How do we evaluate this sum? Well, we note that the sum aS_N is

$$aS_N = \sum_{n=0}^{N-1} a^{n+1} = \sum_{k=1}^{N} a^k$$

$$= \sum_{k=0}^{N-1} a_k + a^N - 1$$ 6.20

$$= S_N + a^N - 1.$$

Therefore, provided $a \neq 1$, the sum S_N is

$$S_N = \frac{1 - a^N}{1 - a}, \qquad a \neq 1. \tag{6.21}$$

This formula, discovered already in Chapter 2, works for $a \neq 1$. When $a = 1$, then $S_N = N$:

$$S_N = \begin{cases} \frac{1-a^N}{1-a}, & a \neq 1 \\ N, & a = 1. \end{cases} \tag{6.22}$$

When $|a| < 1$, then $a^N \to 0$ for $N \to \infty$, and we have the asymptotic formula

$$\lim_{N \to \infty} S_N = \frac{1}{1 - a}, \qquad |a| < 1. \tag{6.23}$$

Problem 6.7 Evaluate $S_N = \sum_{n=0}^{N-1} a^n$ and $X_N = \frac{1}{N} S_N$ for $a = 0.9, 1,$ and 1.1 and for $N = 1, 2, 4, 8, 16,$ and 32. ∎

Problem 6.8 Prove that $S_N = \sum_{n=0}^{N-1} a^n$ obeys the recursion

$$S_N = S_{N-1} + a^{N-1}.$$

Prove that $S_N = N$ obeys this recursion for $a = 1$ and that $S_N = \frac{1-a^N}{1-a}$ obeys it for $a \neq 1$. ∎

Recursive Computation. Every sum of the form

$$S_N = \sum_{n=0}^{N-1} u_n \tag{6.24}$$

obeys the recursion

$$S_N = S_{N-1} + u_{N-1}. \tag{6.25}$$

This means that when summing numbers you may "use them and discard them." That is, you do not need to read them, store them, and sum them.

You may read u_0 to form S_1 and discard u_0; add u_1 to S_1 and discard u_1; add u_2 to S_2; and continue.

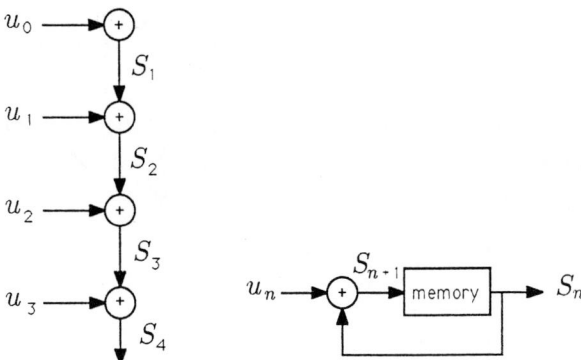

Figure 6.6: The Recursion $S_{n+1} = S_n + u_n$

This is very important for hardware and software implementations of running sums. You need only store the current sum, not the measurements that produced it. Two illustrations of the recursion $S_{n+1} = S_n + u_n$ are provided in Figure 6.6. The diagram on the left is self-explanatory. The diagram on the right says that the sum S_n is stored in a memory location, to be added to u_n to produce S_{n+1}, which is then stored back in the memory location to be added to u_{n+1}, and so on.

6.3 Weighted Averages

Weighted, tapered, or windowed averages are straightforward generalizations of simple averages. They take the form

$$x = \sum_{n=1}^{N} w_n u_n \qquad\qquad 6.26$$

with the constraint that the "weights in the window," w_n, sum to 1:

$$\sum_{n=1}^{N} w_n = 1. \qquad\qquad 6.27$$

When $w_n = \frac{1}{N}$, then x is the simple average studied in Section 6.2.

Example 6.3. There are many windows that are commonly used in engineering practice. For N odd, the standard triangular window is

$$w_n = \frac{2}{N+1}\left(1 - \frac{2}{N+1}\left|\frac{N+1}{2} - n\right|\right). \qquad 6.28$$

This window, illustrated in Figure 6.7, weights the input $u_{(N+1)/2}$ by $\frac{2}{N+1}$ and the inputs u_1 and u_N by $\left(\frac{2}{N+1}\right)^2$. The most general triangular window takes the form

$$w_n = \alpha\left(1 - \beta\left|\frac{N+1}{2} - n\right|\right); \qquad \alpha, \beta > 0, \quad N \text{ odd.} \quad \square \qquad 6.29$$

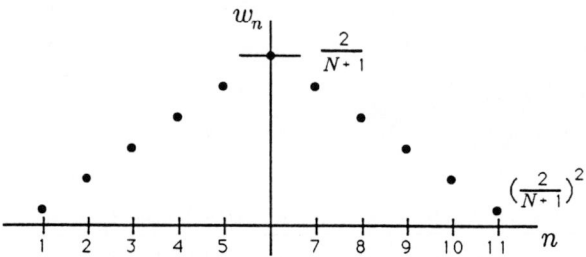

Figure 6.7: Triangular Window

Problem 6.9   Determine the constraints on α and β to make the general triangular window a valid window (i.e., $\sum\limits_{n=1}^{N} w_n = 1$). Show that $\alpha = \frac{2}{N+1} = \beta$ is a valid solution. Propose another solution that you like. ∎

Problem 6.10   You are taking three 3-credit courses, one 5-credit course, and one 2-credit course. Write down the weighted average for computing your GPA in a system that awards 4.0 points for an A, 3.0 points for a B, ... , and (horrors!) 0 points for an F. ∎

6.4 Moving Averages

Moving averages are generalizations of weighted averages. They are designed to "run along an input sequence, computing weighted averages as they go." A typical moving average over N inputs takes the form

$$x_n = \sum_{k=0}^{N-1} w_k u_{n-k} \qquad\qquad 6.30$$

$$= w_0 u_n + w_1 u_{n-1} + \cdots + w_{N-1} u_{n-(N-1)}.$$

The most current input, u_n, is weighted by w_0; the next most current input, u_{n-1}, is weighted by w_1; and so on. This weighting is illustrated in Figure 6.8. The sequence of weights, w_0 through w_{N-1}, is called a "window," a "weighting sequence," or a "filter." In the example illustrated in Figure 6.8, the current value u_n is weighted more heavily than the least current value. This is typical (but not essential) because we usually want x_n to reflect more of the recent past than the distant past.

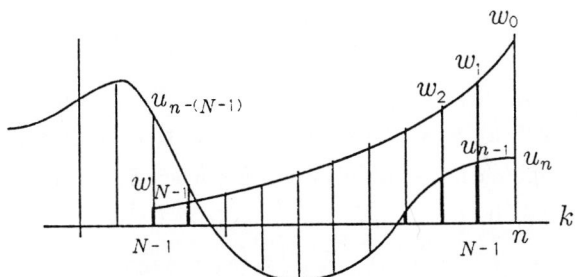

Figure 6.8: Moving Average

Example 6.4. When the weights $w_0, w_1, \ldots, w_{N-1}$ are all equal to $\frac{1}{N}$, then the moving average x_n is a "simple moving average":

$$x_n = \frac{1}{N}[u_n + u_{n-1} + \cdots + u_{N-1}]. \qquad\qquad 6.31$$

This is the same as the simple average that we studied in Section 6.2, but now the simple average moves along the sequence of inputs, averaging the N most current values. □

Problem 6.11 Evaluate the moving average $x_n = \sum_{k=0}^{N-1} \frac{1}{N} u_{n-k}$ for the inputs

(a) $u_n = \begin{cases} 0, & n < 0 \\ u, & n \geq 0; \end{cases}$

(b) $u_n = \begin{cases} 0, & n \leq 0 \\ n, & n > 0. \end{cases}$

Interpret your findings. ∎

Problem 6.12 Evaluate the simple moving average $x_n = \sum_{k=0}^{N-1} \frac{1}{N} u_{n-k}$ when u_n is the sequence

$$u_n = \begin{cases} 0, & n < 0 \\ a^n, & n \geq 0. \end{cases}$$

Interpret your result. ∎

Example 6.5. When the weights w_n equal $w_0 a^n$ for $n = 0, 1, \ldots, N - 1$, then the moving average x_n takes the form

$$x_n = w_0 \sum_{k=0}^{N-1} a^k u_{n-k}. \qquad\qquad 6.32$$

When $a < 1$, then u_n is weighted more heavily than $u_{n-(N-1)}$; when $a > 1$, $u_{n-(N-1)}$ is weighted more heavily than u_n; when $a = 1$, u_n is weighted the same as $u_{n-(N-1)}$. □

Problem 6.13 Evaluate w_0 so that the exponential weighting sequence $w_n = w_0 a^n$ $(n = 0, 1, \ldots, N - 1)$ is a valid window (i.e., $\sum_{n=0}^{N-1} w_n = 1$). ∎

Problem 6.14 Compute the moving average $x_n = \sum_{k=0}^{N-1} w_0 a^k u_{n-k}$ when the input sequence u_n is

$$u_n = \begin{cases} b^n, & n \geq 0 \\ 0, & n < 0. \end{cases}$$

What happens when $b = a$? Can you explain this? ∎

6.5 Exponential Averages and Recursive Filters

Suppose we try to extend our method for computing finite moving averages to infinite moving averages of the form

$$x_n = \sum_{k=0}^{\infty} w_k u_{n-k}$$

6.33

$$= w_0 u_n + w_1 u_{n-1} + \cdots + w_{1000} u_{n-1000} + \cdots .$$

In general, this moving average would require infinite memory for the weighting coefficients $w_0, w_1, \ldots$ and for the inputs $u_n, u_{n-1}, \ldots$. Furthermore, the hardware for multiplying $w_k u_{n-k}$ would have to be infinitely fast to compute the infinite moving average in finite time. All of this is clearly fanciful and implausible (not to mention impossible). But what if the weights take the exponential form

$$w_k = \begin{cases} 0, & k < 0 \\ w_0 a^k, & k \geq 0? \end{cases}$$

6.34

Does any simplification result? There is hope because the weighting sequence obeys the recursion

$$w_k = \begin{cases} 0, & k < 0 \\ w_0, & k = 0 \\ a w_{k-1}, & k \geq 1. \end{cases}$$

6.35

This recursion may be rewritten as follows, for $k \geq 1$:

$$w_k - a w_{k-1} = 0, \qquad k \geq 1.$$

6.36

Let's now manipulate the infinite moving average and use the recursion for the weights to see what happens. You must follow every step:

$$x_n = \sum_{k=0}^{\infty} w_k u_{n-k}$$

$$= \sum_{k=1}^{\infty} w_k u_{n-k} + w_0 u_n$$

$$= \sum_{k=1}^{\infty} a w_{k-1} u_{n-k} + w_0 u_n$$

6.37

$$= a \sum_{m=0}^{\infty} w_m u_{n-1-m} + w_0 u_n$$

$$= a x_{n-1} + w_0 u_n.$$

This result is fundamentally important because it says that the output of the *infinite exponential moving average* may be computed by scaling the previous output x_{n-1} by the constant a, scaling the new input u_n by w_0, and adding. Only three memory locations must be allocated: one for w_0, one for a, and one for x_{n-1}. Only two multiplies must be implemented: one for ax_{n-1} and one for $w_0 u_n$. A diagram of the recursion is given in Figure 6.9. In this recursion, the old value of the exponential moving average, x_{n-1}, is scaled by a and added to $w_0 u_n$ to produce the new exponential moving average x_n. This new value is stored in memory, where it becomes x_{n-1} in the next step of the recursion, and so on.

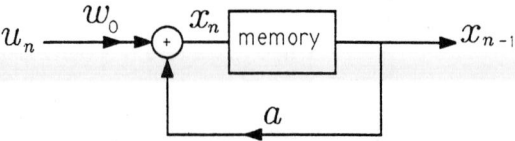

Figure 6.9: Recursive Implementation of an Exponential Moving Average

Problem 6.15 Try to extend the recursion of the previous paragraphs to the weighted average

$$x_n = \sum_{k=0}^{N-1} a^k u_{n-k}.$$

What goes wrong? ∎

Problem 6.16 Compute the output of the exponential moving average $x_n = a x_{n-1} + w_0 u_n$ when the input is

$$u_n = \begin{cases} 0, & n < 0 \\ u, & n \geq 0. \end{cases}$$

Plot your result versus n. ∎

Problem 6.17 Compute w_0 in the exponential weighting sequence

$$w_n = \begin{cases} 0, & n < 0 \\ a^n w_0, & n \geq 0 \end{cases}$$

to make the weighting sequence a valid window. (This is a special case of Problem 6.13.) Assume $-1 < a < 1$. ∎

6.6 Test Sequences

When we design a filter, we design it for a purpose. For example, a moving average filter is often designed to pass relatively constant data while averaging out relatively variable data. In an effort to clarify the behavior of a filter, we typically analyze its response to a standard set of test signals. We will call the *impulse*, the *step*, and the *complex exponential* the standard test signals.

Unit Pulse Sequence. The unit pulse sequence is the sequence

$$u_n = \delta_n = \begin{cases} 1, & n = 0 \\ 0, & n \neq 0. \end{cases} \qquad 6.38$$

This sequence, illustrated in Figure 6.10, consists of all zeros except for a single one at $n = 0$. If the unit pulse sequence is passed through a moving average filter (whether finite or not), then the output is called the *unit pulse response*:

$$h_n = \sum_{k=0}^{\infty} w_k \delta_{n-k} \qquad 6.39$$

$$= w_n.$$

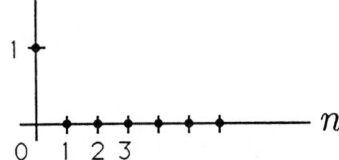

Figure 6.10: Unit Pulse Sequence

(Note that $\delta_{n-k} = 0$ unless $n = k$.) So the unit pulse sequence may be used to read out the weights of a moving average filter. It is common practice to use w_k (the kth weight) and h_k (the kth impulse response) interchangeably.

Problem 6.18   Find the unit pulse response for the finite moving average $x_n = \sum\limits_{k=0}^{N-1} w_k u_{n-k}$. Caution: You must consider $n < 0$, $0 \le n \le N - 1$, and $n \ge N$. ∎

Problem 6.19   Find the unit pulse response for the recursive filter $x_n = a x_{n-1} + w_0 u_n$. ∎

Unit Step Sequence. The unit step sequence is the sequence

$$u_n = \xi_n = \begin{cases} 1, & n \ge 0 \\ 0, & n < 0. \end{cases} \tag{6.40}$$

This sequence is illustrated in Figure 6.11. When this sequence is applied to a moving average filter, the result is the *unit step response*

$$g_n = \sum_{k=0}^{n} w_k$$
$$= \sum_{k=0}^{n} h_k. \tag{6.41}$$

The unit step response is just the sequence of partial sums of the unit pulse response.

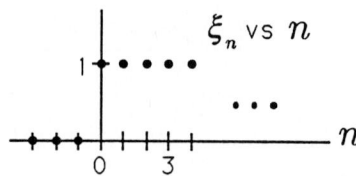

Figure 6.11: Unit Step Sequence

Problem 6.20   Find the unit step response for the finite moving average filter $x_n = \sum\limits_{k=0}^{N-1} w_k u_{n-k}$. Specialize your general result to the special case where $w_k = \frac{1}{N}$ for $k = 0, 1, \ldots, N - 1$. ∎

Problem 6.21   Find the unit step response for the recursive filter $x_n = a x_{n-1} + w_0 u_n$. ∎

Complex Exponential Sequence. The complex exponential sequence is the sequence

$$u_k = e^{jk\theta}, \qquad k = 0, \pm1, \pm2, \dots .\qquad 6.42$$

This sequence, illustrated in Figure 6.12, is a "discrete-time phasor" that "ratchets" counterclockwise (CCW) as k moves to $k+1$ and clockwise (CW) as k moves to $k-1$. Each time the phasor ratchets, it turns out an angle of θ. Why should such a sequence be a useful test sequence? There are two reasons.

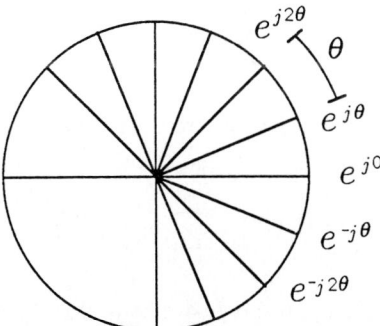

Figure 6.12: Discrete-Time Phasor

(i) $e^{jk\theta}$ **represents (or codes)** $\cos k\theta$. The real part of the sequence $e^{jk\theta}$ is the cosinusoidal sequence $\cos k\theta$:

$$\text{Re}[e^{jk\theta}] = \cos k\theta.\qquad 6.43$$

Therefore the discrete-time phasor $e^{jk\theta}$ represents (or codes) $\cos k\theta$ in the same way that the continuous-time phasor $e^{j\omega t}$ codes $\cos \omega t$. If the moving average filter

$$x_n = \sum_{k=0}^{\infty} h_k u_{n-k}\qquad 6.44$$

has real coefficients, we can get the response to a cosinusoidal sequence by taking the real part of the following sum:

$$x_n = \sum_{k=0}^{\infty} h_k \cos(n-k)\theta$$

$$= \operatorname{Re}\left[\sum_{k=0}^{\infty} h_k e^{j(n-k)\theta}\right] \qquad 6.45$$

$$= \operatorname{Re}\left[e^{jn\theta} \sum_{k=0}^{\infty} h_k e^{-jk\theta}\right].$$

In this formula, the sum

$$\sum_{k=0}^{\infty} h_k e^{-jk\theta} \qquad 6.46$$

is called the *complex frequency response* of the filter and is given the symbol

$$H(e^{j\theta}) = \sum_{k=0}^{\infty} h_k e^{-jk\theta}. \qquad 6.47$$

This complex frequency response is just a complex number, with a magnitude $|H(e^{j\theta})|$ and a phase $\arg H(e^{j\theta})$. Therefore the output of the moving average filter is

$$x_n = \operatorname{Re}\left[e^{jn\theta} H(e^{j\theta})\right]$$

$$= \operatorname{Re}\left[e^{jn\theta} |H(e^{j\theta})| e^{j \arg H(e^{j\theta})}\right] \qquad 6.48$$

$$= |H(e^{j\theta})| \cos\left[n\theta + \arg H(e^{j\theta})\right].$$

This remarkable result says that the output is also cosinusoidal, but its amplitude is $|H(e^{j\theta})|$ rather than 1, and its phase is $\arg H(e^{j\theta})$ rather than 0. In the examples to follow, we will show that the complex "gain" $H(e^{j\theta})$ can be highly selective in θ, meaning that cosines of some angular frequencies are passed with little attenuation while cosines of other frequencies are dramatically attenuated. By choosing the filter coefficients, we can design the frequency selectivity we would like to have.

(ii) $e^{jk\theta}$ **is a sampled data version of** $e^{j\omega t}$. The discrete-time phasor $e^{jk\theta}$ can be produced physically by sampling the continuous-time phasor $e^{j\omega t}$ at the periodic sampling instants $t_k = kT$:

$$e^{jk\theta} = e^{j\omega t}\Big|_{t=kT} = e^{j\omega kT} \qquad\qquad 6.49$$

$$\theta = \omega T.$$

The dimensions of θ are radians, the dimensions of ω are radians/second, and the dimensions of T are seconds. We call T the sampling interval and $\frac{1}{T}$ the sampling rate or sampling frequency. If the original angular frequency of the phasor $e^{j\omega t}$ is increased to $\omega + m\left(\frac{2\pi}{T}\right)$, then the discrete-time phasor remains $e^{jk\theta}$:

$$e^{j[\omega+m(2\pi/T)]t}\Big|_{t=kT} = e^{j(\omega kT+km2\pi)} = e^{jk\theta}. \qquad\qquad 6.50$$

This means that all continuous-time phasors of the form $e^{j[\omega+m(2\pi/T)]t}$ "hide under the same alias" when viewed through the sampling operation. That is, the sampled-data phasor cannot distinguish the frequency ω from the frequency $\omega + m\frac{2\pi}{T}$. In your subsequent courses you will study aliasing in more detail and study the Nyquist rule for sampling:

$$T \le \frac{2\pi}{\Omega}; \qquad \frac{1}{T} \ge \frac{\Omega}{2\pi}. \qquad\qquad 6.51$$

This rule says that you must sample signals at a rate $\left(\frac{1}{T}\right)$ that exceeds the *bandwidth* $\frac{\Omega}{2\pi}$ of the signal.

Example 6.6. Let's pass the cosinusoidal sequence $u_k = \cos k\theta$ through the finite moving average filter

$$x_n = \sum_{k=0}^{N-1} h_k u_{n-k} \qquad\qquad 6.52$$

$$h_k = \frac{1}{N}, \qquad k = 0, 1, \ldots, N-1.$$

We know from our previous result that the output is

$$x_n = \left|H(e^{j\theta})\right| \cos\left[n\theta + \arg H(e^{j\theta})\right]. \qquad\qquad 6.53$$

The complex frequency response for this example is

$$H(e^{j\theta}) = \sum_{k=0}^{N-1} \frac{1}{N} e^{-jk\theta}$$

$$= \frac{1}{N} \frac{1 - e^{-jN\theta}}{1 - e^{-j\theta}}.$$

6.54

(Do you see your old friend, the finite sum formula, at work?) Let's try to manipulate the result into a more elegant form:

$$H(e^{j\theta}) = \frac{1}{N} \frac{e^{-j(N/2)\theta}[e^{j(N/2)\theta} - e^{-j(N/2)\theta}]}{e^{-j(\theta/2)}[e^{j(\theta/2)} - e^{-j(\theta/2)}]}$$

$$= \frac{1}{N} e^{-j[(N-1)/2]\theta} \frac{\sin\left(\frac{N}{2}\theta\right)}{\sin\left(\frac{1}{2}\theta\right)}.$$

6.55

The magnitude of the function $H(e^{j\theta})$ is

$$|H(e^{j\theta})| = \frac{1}{N} \left| \frac{\sin\left(\frac{N}{2}\theta\right)}{\sin\left(\frac{1}{2}\theta\right)} \right|.$$

6.56

At $\theta = 0$, corresponding to a "DC phasor," $H(e^{j\theta})$ equals 1; at $\theta = \frac{2\pi}{N}$, $|H(e^{j\theta})| = 0$. The magnitude of the complex frequency response is plotted in Figure 6.13.

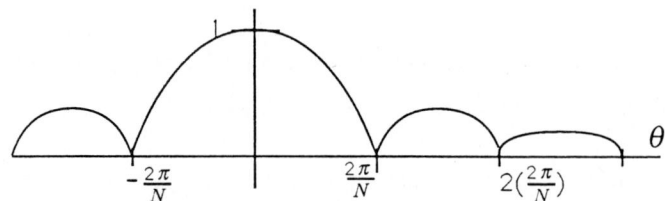

Figure 6.13: Frequency Selectivity of a Moving Average Filter

This result shows that the moving average filter is frequency selective, passing low frequencies with gain near 1 and high frequencies with gain near 0. □

Problem 6.22 Compute the phase of the complex frequency response

$$H(e^{j\theta}) = \frac{1}{N} e^{-j[(N-1)/2]\theta} \frac{\sin\left(\frac{N}{2}\theta\right)}{\sin\left(\frac{1}{2}\theta\right)}. \quad \blacksquare$$

Problem 6.23 Choose the filter length N for the filter $h_k = \frac{1}{N}$, $k = 0, 1, \ldots, N-1$, so that a 60 Hz cosine, sampled at the rate $\frac{1}{T} = 180$, is perfectly zeroed out as it comes through the filter. ∎

Problem 6.24 (MATLAB) Write a MATLAB program to compute and plot the magnitude $\left| H(e^{j\theta}) \right|$ and the phase $\arg H(e^{j\theta})$ versus $-\pi < \theta < \pi$ when

$$H(e^{j\theta}) = \frac{1}{N} e^{-j[(N-1)/2]\theta} \frac{\sin \left(\frac{N}{2} \theta \right)}{\sin \left(\frac{1}{2} \theta \right)}.$$

Choose suitable increments for θ. ∎

Problem 6.25 Compute the complex frequency response $H(e^{j\theta})$ for the recursive filter $x_n = ax_{n-1} + w_0 u_n$. ∎

6.7 Numerical Experiment (Frequency Response of First-Order Filter)

Consider the exponential moving average filter

$$x_n = \sum_{k=0}^{\infty} a^k u_{n-k}; \qquad a = 0.98. \qquad\qquad 6.57$$

(1) Write out a few terms of the sum to show how the filter works.

(2) Write x_n as a recursion and discuss the computer memory required to implement the filter.

(3) Compute the complex frequency response $H(e^{j\theta})$ for the filter.

(4) Write a MATLAB program to plot the magnitude and phase of the complex frequency response $H(e^{j\theta})$ versus θ for $\theta = -\pi$ to $+\pi$ in steps of $\frac{2\pi}{64}$. Do this for two values of a, namely, $a = 0.98$ and $a = -0.98$. Explain your findings.

(5) Write a MATLAB program to pass the following signals through the filter when $a = 0.98$:

  (a) $u_n = \delta_n$

  (b) $u_n = \xi_n$

(c) $u_n = \xi_n \cos \frac{2\pi}{64} n$

(d) $u_n = \xi_n \cos \frac{2\pi}{32} n$

(e) $u_n = \xi_n \cos \frac{2\pi}{16} n$

(f) $u_n = \xi_n \cos \frac{2\pi}{8} n$

(g) $u_n = \xi_n \cos \frac{2\pi}{4} n$

(h) $u_n = \xi_n \cos \frac{2\pi}{2} n.$

Plot the outputs for each case and interpret your findings in terms of the complex frequency response $H(e^{j\theta})$. Repeat (5) for $a = -0.98$. Interpret your findings.

7

Binary Codes

Notes to Teachers and Students:

We use this chapter to introduce students to the communication paradigm and to show how arbitrary symbols may be represented by binary codes. These symbols and their corresponding binary codes may be computer instructions, integer data, approximations to real data, and so on.

We develop some ad hoc tree codes for representing information and then develop Huffman codes for optimizing the use of bits. Hamming codes add check bits to a binary word so that errors may be detected and corrected. The numerical experiment in Section 7.6 has the students design a Huffman code for coding Lincoln's Gettysburg Address.

7.1 Introduction

It would be stretching our imagination to suggest that Sir Francis had digital audio on his minde (sic) when he wrote the prophetic words

> . . . a man may expresse and signifie the intentions of his minde, at any distance . . . by . . . objects . . . capable of a twofold difference onely.
>
> –Sir Francis Bacon, 1623

Nonetheless, this basic idea forms the basis of everything we do in digital computing, digital communications, and digital audio/video. In 1832, Samuel F. B. Morse used the very same idea to propose that telegram *words* be coded into *binary addresses* or *binary codes* that could be transmitted over telegraph lines and decoded at the receiving end to unravel the telegram. Morse abandoned his scheme, illustrated in Figure 7.1, as too complicated and, in 1838, proposed his fabled Morse code for coding letters (instead of words) into objects (dots, dashes, spaces) capable of a threefold difference onely (sic).

Acknowledgment: Richard Hamming's book, *Information Theory and Coding*, Prentice-Hall, New York (1985) and C. T. Mullis's unpublished notes have influenced our treatment of binary codes. The numerical experiment was developed by Mullis.

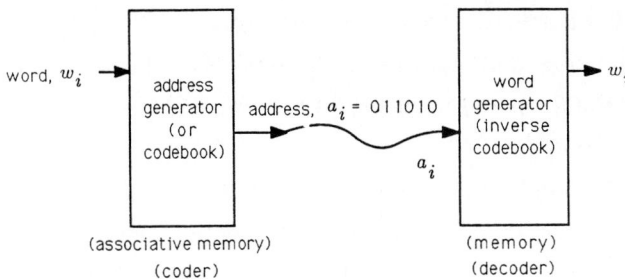

Figure 7.1: Generalized Coder-Decoder

The basic idea of Figure 7.1 is used today in cryptographic systems, where the "address a_i" is an encyphered version of a message w_i; in vector quantizers, where the "address a_i" is the address of a close approximation to data w_i; in coded satellite transmissions, where the "address a_i" is a data word w_i plus parity check bits for detecting and correcting errors; in digital audio systems, where the "address a_i" is a stretch of digitized and coded music; and in computer memories, where a_i is an address (a coded version of a word of memory) and w_i is a word in memory.

In this chapter we study three fundamental questions in the construction of binary addresses or binary codes. First, what are plausible schemes for mapping symbols (such as words, letters, computer instructions, voltages, pressures, etc.) into binary codes? Second, what are plausible schemes for coding likely symbols with short binary words and unlikely symbols with long words in order to minimize the number of binary digits (bits) required to represent a message? Third, what are plausible schemes for "coding" binary words into longer binary words that contain "redundant bits" that may be used to detect and correct errors? These are not new questions. They have occupied the minds of many great thinkers. Sir Francis recognized that arbitrary messages had binary representations. Alan Turing, Alonzo Church, and Kurt Goedel studied binary codes for computations in their study of computable numbers and algorithms. Claude Shannon, R. C. Bose, Irving Reed, Richard Hamming, and many others have studied error control codes. Shannon, David Huffman, and many others have studied the problem of efficiently coding information.

In this chapter we outline the main ideas in binary coding and illustrate the role that binary coding plays in digital communications. In your subsequent courses in electrical and computer engineering you will study integrated circuits for building coders and decoders and mathematical models for designing good codes.

7.2 The Communication Paradigm

A *paradigm* is a pattern of ideas that form the foundation for a body of knowledge. A paradigm for (tele-) communication theory is a pattern of basic building blocks that may be applied to the dual problems of (i) reliably transmitting information from source to receiver at high speed or (ii) reliably storing information from source to memory at high density. High-speed communication permits us to accommodate many low-rate sources (such as audio) or one high-rate source (such as video). High-density storage permits us to store a large amount of information in a small space. For example, a typical 1.2 Mbyte floppy disc stores 9.6×10^6 bits of information, whereas a typical CD stores about 2×10^9 bits, enough for one hour's worth of high-quality sound.

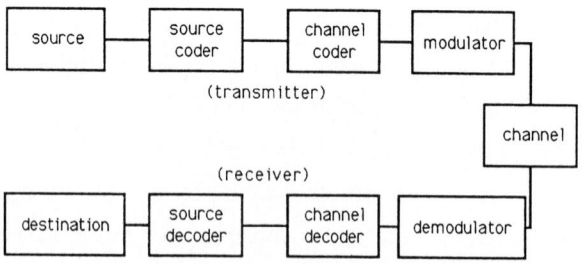

Figure 7.2: Basic Building Blocks in a (Tele-) Communication System

Figure 7.2 illustrates the basic building blocks that apply to any problem in the theory of (tele-) communication. The *source* is an arbitrary source of information. It can be the time-varying voltage at the output of a vibration sensor (such as an integrating accelerometer for measuring motion or a microphone for measuring sound pressure); it can be the charges stored in the CCD array of a solid-state camera; it can be the addresses generated from a sequence of keystrokes at a computer terminal; it can be a sequence of instructions in a computer program. The *source coder* is a device for turning primitive source outputs into more efficient representations. For example, in a recording studio, the source coder would convert analog voltages into digital approximations using an A/D converter; a fancy source coder would use a fancy A/D converter that finely quantized likely analog values and crudely quantized unlikely values. If the source is a source of discrete symbols like letters and numbers, then a fancy source code would assign short binary sequences to likely symbols (such as e) and long binary sequences to unlikely symbols (such as z). The *channel coder* adds "redundant bits" to the binary output of the source coder so that errors of transmission or storage may be detected and corrected. In the simplest example, a binary string of the form 01001001 would have an extra bit of 1 added to give *even parity* (an even number of 1's) to the string; the string 10110111 would have an extra bit of 0 added to preserve the *even parity*. If one bit error is introduced in the channel, then the parity is odd and the receiver knows that an error has occurred. The *modulator* takes outputs of the channel coder, a stream of binary digits, and constructs an analog waveform that represents a block of bits. For example, in a 9600 baud Modem, five bits are used to determine one of $2^5 = 32$ phases that are used to modulate the signal $A\cos(\omega t + \phi)$. Each possible string of five bits has its own personalized phase, ϕ, and this phase can be determined at the receiver. The signal $A\cos(\omega t + \phi)$ is an analog signal that may be transmitted over a *channel* (such as a telephone line, a microwave link, or a fiber-optic cable). The channel has a finite bandwidth, meaning that it distorts signals, and it is subject to noise or interference from other electromagnetic radiation. Therefore transmitted information arrives at the *demodulator* in imperfect form. The demodulator uses *filters* matched to the modulated signals to demodulate the phase and look up the corresponding bit stream. The *channel decoder* converts the coded bit stream into the information bit stream, and the *source decoder* looks up the corresponding symbol. This sequence of steps is illustrated symbolically in Figure 7.3.

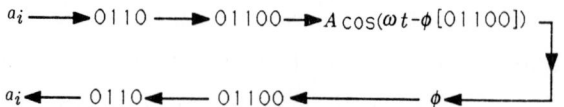

Figure 7.3: Symbolic Representation of Communication

In your subsequent courses on communication theory you will study each block of Figure 7.2 in detail. You will find that every source of information has a characteristic complexity, called *entropy*, that determines the minimum rate at which bits must be generated in order to represent the source. You will also find that every communication channel has a characteristic tolerance for bits, called *channel capacity*. This capacity depends on signal-to-noise ratio and bandwidth. When the channel capacity exceeds the source entropy, then you can transmit information reliably; if it does not, then you cannot.

7.3 From Symbols to Binary Codes

Perhaps the most fundamental idea in communication theory is that arbitrary symbols may be represented by strings of binary digits. These strings are called binary words, binary addresses, or binary codes. In the simplest of cases, a finite alphabet consisting of the letters or symbols $s_0, s_1, \ldots, s_{M-1}$ is represented by binary codes. The obvious way to implement the representation is to let the ith binary code be the binary representation for the subscript i:

$$s_0 \sim 000 = a_0$$

$$s_1 \sim 001 = a_1$$

$$\vdots$$

$$s_6 \sim 110 = a_6$$

$$s_7 \sim 111 = a_7.$$

7.1

The number of bits required for the binary code is N where

$$2^{N-1} < M \leq 2^N.$$

7.2

We say, roughly, that $N = \log_2 M$.

Octal Codes. When the number of symbols is large and the corresponding binary codes contain many bits, then we typically group the bits into groups of three and replace the binary code by its corresponding octal code. For example, a seven-bit binary code maps into a three-digit octal code as follows:

$$0000000 \sim 000$$

$$0000001 \sim 001$$

$$\vdots$$

$$0100110 \sim 046$$

$$\vdots$$

$$101111 \sim 137$$

$$\vdots$$

$$1111111 \sim 177.$$

7.3

The octal ASCII codes for representing letters, numbers, and special characters are tabulated in Figure 7.4.

Problem 7.1 Write out the seven-bit ASCII codes for A, q, 7, and { . ∎

	'0	'1	'2	'3	'4	'5	'6	'7
'00x	NUL	SOH	STX	ETX	EOT	ENQ	ACK	BEL
'01x	BS	HT	LF	VT	FF	CR	SO	SI
'02x	DLE	DC1	DC2	DC3	DC4	NAK	SYN	ETB
'03x	CAN	EM	SUB	ESC	FS	GS	RS	US
'04x	SP	!	"	#	$	%	&	'
'05x	(	)	*	+	,	−	.	/
'06x	0	1	2	3	4	5	6	7
'07x	8	9	:	;	<	=	>	?
'10x	@	A	B	C	D	E	F	G
'11x	H	I	J	K	L	M	N	O
'12x	P	Q	R	S	T	U	V	W
'13x	X	Y	Z	[	\	]	^	_
'14x	`	a	b	c	d	e	f	g
'15x	h	i	j	k	l	m	n	o
'16x	p	q	r	s	t	u	v	w
'17x	x	y	z	{	|	}	~	DEL

Figure 7.4: Octal ASCII Codes (from Donald E. Knuth, *The TEXbook*, ©1986 by the American Mathematical Society, Providence, Rhode Island p. 367, published by Addison-Wesley Publishing Co.)

Problem 7.2   Add a 1 or a 0 to the most significant (left-most) position of the seven-bit ASCII code to produce an eight-bit code that has even parity (even number of 1's). Give the resulting eight-bit ASCII codes and the corresponding three-digit octal codes for %, u, f, 8, and +. ∎

Quantizers and A/D Converters. What if the source alphabet is infinite? Our only hope is to approximate it with a finite collection of finite binary words. For example, suppose the output of the source is an analog voltage that lies between $-V_0$ and $+V_0$. We might break this peak-to-peak range up into little voltage cells of size $\frac{2V_0}{M}$ and approximate the voltage in each cell by its midpoint. This scheme is illustrated in Figure 7.5. In the figure, the cell C_i is defined to be the set of voltages that fall between $i\frac{2V_0}{M} - \frac{V_0}{M}$ and $i\frac{2V_0}{M} + \frac{V_0}{M}$:

$$C_i = \left\{ V : i\frac{2V_0}{M} - \frac{V_0}{M} < V \le i\frac{2V_0}{M} + \frac{V_0}{M} \right\}. \qquad 7.4$$

The mapping from continuous values of V to a finite set of approximations is

$$Q(V) = i\frac{2V_0}{M}, \qquad \text{if } V \in C_i. \qquad 7.5$$

That is, V is replaced by the quantized approximation $i\frac{2V_0}{M}$ whenever V lies in cell C_i. We may represent the quantized values $i\frac{2V_0}{M}$ with binary codes by simply representing the subscript of the cell by a binary word. In a subsequent course on digital electronics and microprocessors you will study A/D (analog-to-digital) converters for quantizing variables.

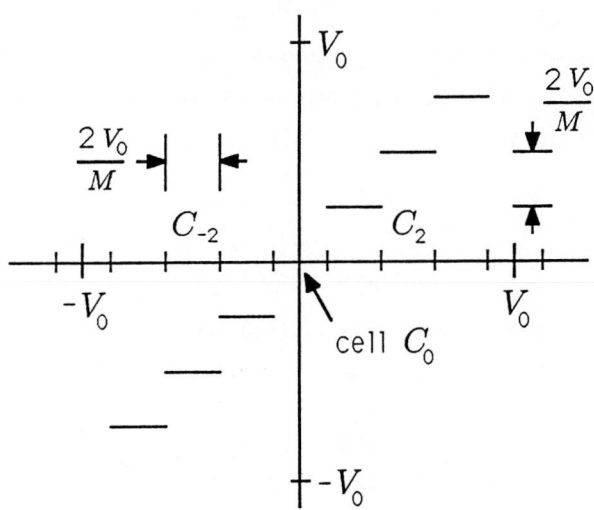

Figure 7.5: A Quantizer

Example 7.1. If $M = 8$, corresponding to a three-bit quantizer, we may associate quantizer cells and quantized levels with binary codes as follows:

$$V \in C_{-3} \implies V_{-3} = (-3)\frac{2V_0}{8} \sim 111$$

$$V \in C_{-2} \implies V_{-2} = (-2)\frac{2V_0}{8} \sim 110$$

$$V \in C_{-1} \implies V_{-1} = (-1)\frac{2V_0}{8} \sim 101$$

$$V \in C_0 \implies V_0 = 0 \sim 000 \qquad\qquad 7.6$$

$$V \in C_1 \implies V_1 = (1)\frac{2V_0}{8} \sim 001$$

$$V \in C_2 \implies V_2 = (2)\frac{2V_0}{8} \sim 010$$

$$V \in C_3 \implies V_3 = (3)\frac{2V_0}{8} \sim 011.$$

This particular code is called a *sign-magnitude code*, wherein the most significant bit is a sign bit and the remaining bits are magnitude bits (e.g., $110 \sim -2$ and $010 \sim 2$). One of the defects of the sign-magnitude code is that it wastes one code by using 000 for 0 and 100 for -0. An alternative code that has many other advantages is the *2's complement code*. The 2's complement codes for positive numbers are the same as the sign-magnitude codes, but the codes for negative numbers are generated by complementing all bits for the corresponding positive number and adding 1:

$$-4 \sim 100$$

$$-3 \sim 101 \qquad (100 + 1)$$

$$-2 \sim 110 \qquad (101 + 1)$$

$$-1 \sim 111 \qquad (110 + 1)$$

$$\qquad\qquad\qquad\qquad 7.7$$

$$0 \sim 000$$

$$1 \sim 001$$

$$2 \sim 010$$

$$3 \sim 011. \qquad\qquad \square$$

Problem 7.3 Generate the four-bit sign-magnitude and four-bit 2's complement binary codes for the numbers $-8, -7, \ldots, -1, 0, 1, 2, \ldots, 7$. ∎

Problem 7.4   Prove that, in the 2's complement representation, the binary codes for $-n$ and $+n$ sum to zero. For example,

$$101 + 011 = 000$$
$$(-3) \quad (3) \quad (0). \quad \blacksquare$$

In your courses on computer arithmetic you will learn how to do arithmetic in various binary-coded systems. The following problem illustrates how easy arithmetic is in 2's complement.

Problem 7.5   Generate a table of sums for all 2's complement numbers between -4 and $+3$. Show that the sums are correct. Use $0 + 0 = 0$, $0 + 1 = 1, 1 + 0 = 1$, and $1 + 1 = 0$ with a carry into the next bit. For example, $001 + 001 = 010$. $\blacksquare$

Binary Trees and Variable-Length Codes.   The codes we have constructed so far are constant-length codes for finite alphabets that contain exactly $M = 2^N$ symbols. In the case where $M = 8$ and $N = 3$, then the eight possible three-bit codes may be represented as leaves on the branching tree illustrated in Figure 7.6(a). The tree grows a left branch for a 0 and a right branch for a 1, until it terminates after three branchings. The three-bit codes we have studied so far reside at the terminating leaves of the binary tree. But what if our source alphabet contains just five symbols or letters? We can represent these five symbols as the three-bit symbols 000 through 100 on the binary tree. This generates a constant-length code with three unused, or illegal, symbols 101 through 111. These are marked with an "x" in Figure 7.6(a). These unused leaves and the branches leading to them may be pruned to produce the binary tree of Figure 7.6(b).

If we admit variable-length codes, then we have several other options for using a binary tree to construct binary codes. Two of these codes and their corresponding binary trees are illustrated in Figure 7.7. If we disabuse ourselves of the notion that each code word must contain three or fewer bits, then we may construct binary trees like those of Figure 7.8 and generate their corresponding binary codes. In Figure 7.8(a), we grow a right branch after each left branch and label each leaf with a code word. In Figure 7.8(b), we prune off the last right branch and associate a code word with the leaf on the last left branch.

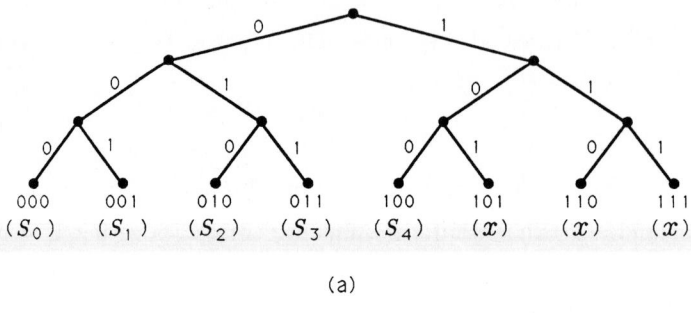

(a)

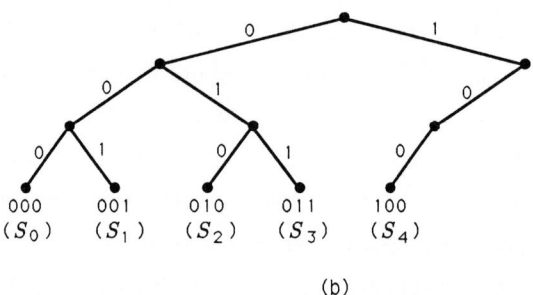

(b)

Figure 7.6: Binary Trees and Constant-Length Codes; (a) Binary Tree, and (b) Pruned Binary Tree

All of the codes we have generated so far are organized in Table 7.1. For each code, the average number of bits/symbol is tabulated. This average ranges from 2.4 to 3.0. If all symbols are equally likely to appear, then the best variable-length code would be code 2.

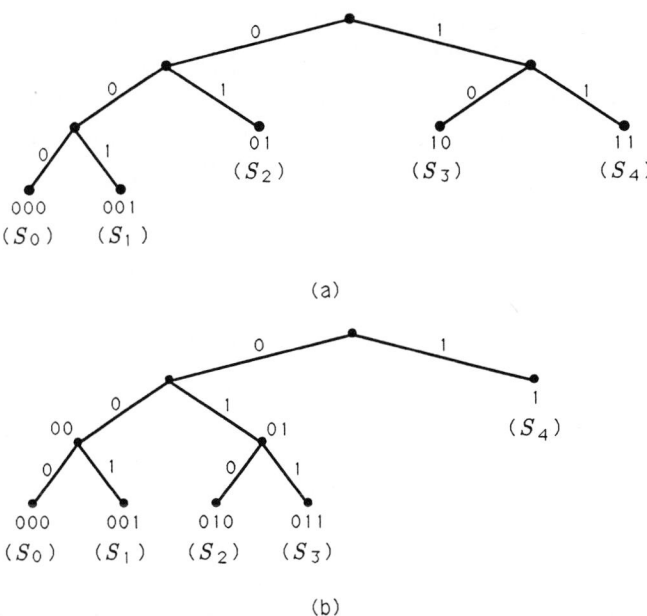

Figure 7.7: Binary Trees and Variable-Length Codes; (a) Binary Tree for Variable-length Code, and (b) Another Binary Tree for Variable-length Code

All of the codes we have constructed have a common characteristic: each code word is a terminating leaf on a binary tree, meaning that no code word lies along a limb of branches to another code word. We say that no code word is a *prefix* to another code word. This property makes each of the codes *instantaneously decodable*, meaning that each bit in a string of bits may be processed instantaneously (or independently) without dependence on subsequent bits.

Problem 7.6 Decode the following sequence of bits using code 2:

01110011110000001011001111. ∎

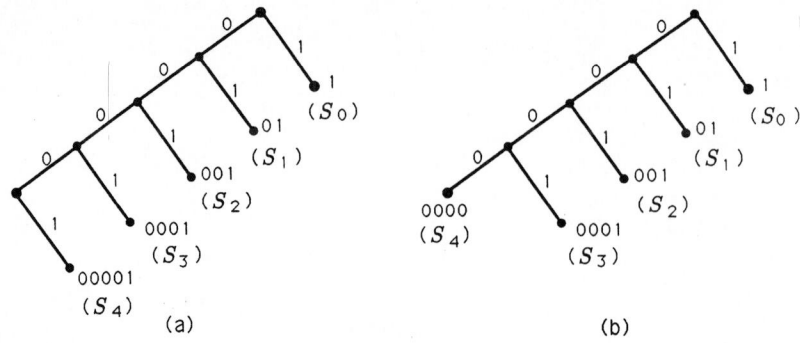

Figure 7.8: Left-Handed Binary Trees for Variable-Length Codes; (a) Left-handed Binary Tree, and (b) Pruned Binary Tree

Table 7.1: Variable Length Codes

Code #	S_0	S_1	S_2	S_3	S_4	Average Bits/Symbol
1	000	001	010	011	100	$15/5 = 3.0$
2	000	001	01	10	11	$12/5 = 2.4$
3	000	001	010	011	1	$13/5 = 2.6$
4	1	01	001	0001	00001	$15/5 = 3.0$
5	1	01	001	0001	0000	$14/5 = 2.8$

Problem 7.7 Illustrate the following codes on a binary tree. Which of them are instantaneously decodable? Which can be pruned and remain instantaneously decodable?

S_0	S_1	S_2	S_3	S_4
011	100	00	11	101
011	100	00	0	01
010	000	100	101	111 . ∎

Code #2 generated in Table 7.1 seems like a better code than code #5 because its average number of bits/symbol (2.4) is smaller. But what if symbol S_0 is a very likely symbol and symbol S_4 is a very unlikely one? Then it may well turn out that the average number of bits used by code #5 is less than the average number used by code #2. So what is the best code? The answer depends on the relative frequency of use for each symbol. We explore this question in the next section.

7.4 Huffman Codes for Source Coding

In 1838, Samuel Morse was struggling with the problem of designing an efficient code for transmitting information over telegraph lines. He reasoned that an efficient code would use short code words for common letters and long code words for uncommon letters. (Can you see the profit motive at work?) In order to turn this reasoned principle into workable practice, Morse rummaged around in the composition trays for typeface in a printshop. He discovered that typesetters use many more e's than z's. He then formed a table that showed the relative frequency with which each letter was used. His ingenious, variable-length Morse code assigned short codes to likely letters (like "dot" for "e") and long codes to unlikely letters (like "dash dash dot dot" for "z"). We now know that Morse came within about 15% of the theoretical minimum for the average code word length for English language text.

A Variation on Morse's Experiment. In order to set the stage for our study of efficient source codes, let's run a variation on Morse's experiment to see if we can independently arrive at a way of designing codes. Instead of giving ourselves a composition tray, let's start with a communication source that generates five symbols or letters S_0, S_1, S_2, S_3, S_4. We run the source for 100 transmissions and observe the following numbers of transmissions for each symbol:

$$50 \ S_0\text{'s}$$

$$20 \ S_1\text{'s}$$

$$20 \ S_2\text{'s} \qquad\qquad\qquad 7.8$$

$$5 \ S_3\text{'s}$$

$$5 \ S_4\text{'s}.$$

We will assume that these "source statistics" are typical, meaning that 1000 transmissions would yield 500 S_0's and so on.

The most primitive binary code we could build for our source would use three bits for each symbol:

$$S_0 \sim 000$$

$$S_1 \sim 001$$

$$S_2 \sim 010$$

$$S_3 \sim 011$$

$$S_4 \sim 100 \qquad\qquad\qquad 7.9$$

$$x \sim 101$$

$$x \sim 110$$

$$x \sim 111.$$

This code is inefficient in two ways. First, it leaves three illegal code words that correspond to no source symbol. Second, it uses the same code word length for an unlikely symbol (like S_4) that it uses for a likely symbol (like S_0). The first defect we can correct by concatenating consecutive symbols into symbol blocks, or composite symbols. If we form a composite symbol consisting of M source symbols, then a typical composite symbol is $S_1 S_0 S_1 S_4 S_2 S_3 S_1 S_2 S_0$. The number of such composite symbols that can be generated is 5^M. The binary code for these 5^M composite symbols must contain N binary digits where

$$2^{N-1} < 5^M < 2^N \qquad (N \cong M \log_2 5). \qquad\qquad 7.10$$

The number of bits per source symbol is

$$\frac{N}{M} \cong \log_2 5 = 2.32. \qquad\qquad 7.11$$

This scheme improves on the best variable length code of Table 7.1 by 0.08 bits/symbol.

Problem 7.8 Suppose your source of information generates the 26 lowercase roman letters used in English language text. These letters are to be concatenated into blocks of length M. Complete the following table of N (number of bits) versus M (number of letters in a block) and show that $\frac{N}{M}$ approaches $\log_2 26$.

		M				
	1	2	3	4	5	6
N	5	10				
N/M	5	5				∎

Now let's reason, as Morse did, that an efficient code would use short codes for likely symbols and long codes for unlikely symbols. Let's pick code #5 from Table 7.1 for this purpose:

S_0	S_1	S_2	S_3	S_4
1	01	001	0001	0000 .

This is a variable-length code. If we use this code on the 100 symbols that generated our source statistic, the average number of bits/symbol is

$$\frac{1}{100}\left[50(1) + 20(2) + 20(3) + 5(4) + 5(4)\right] = 1.90 \text{ bits/symbol.}$$

Problem 7.9 Use the source statistics of Equation 7.8 to determine the average number of bits/symbol for each code in Table 7.1. ∎

Entropy. So far, each ad hoc scheme we have tried has produced an improvement in the average number of bits/symbol. How far can this go? The answer is given by Shannon's source coding theorem, which says that the minimum number of bits/symbol is

$$\frac{N}{M} \geq -\sum_{i=1}^{M} p_i \log_2 p_i \qquad 7.12$$

where p_i is the probability that symbol S_i is generated and $-\sum p_i \log_2 p_i$ is a fundamental property of the source called *entropy*. For our five-symbol example, the table of p_i and $-\log p_i$ is given in Table 7.2. The entropy is 1.861, and the bound on bits/symbol is

$$\frac{N}{M} \geq 1.861. \qquad 7.13$$

Code #5 comes within 0.039 of this lower bound. As we will see in the next paragraphs, this is as close as we can come without coding composite symbols.

Table 7.2: Source Statistics for Five-Symbol Source

Symbol	Probability	− Log Probability
S_0	0.5	1
S_1	0.2	2.32
S_2	0.2	2.32
S_3	0.05	4.32
S_4	0.05	4.32

Problem 7.10 Select an arbitrary page of English text. Build a table of source statistics containing p_i (relative frequencies) and $-\log p_i$ for a through z. (Ignore distinction between upper and lower case and ignore punctuation and other special symbols.) Compute the entropy $-\sum_{i=1}^{26} p_i \log_2 p_i$. ∎

Huffman Codes. In the late 1950s, David Huffman discovered an algorithm for designing variable-length codes that minimize the average number of bits/symbol. Huffman's algorithm uses a principle of optimality that says, "the optimal code for M letters has imbedded in it the optimal code for the $M - 1$ letters that result from aggregating the two least likely symbols." When this principle is iterated, then we have an algorithm for generating the binary tree for a Huffman code:

(i) label all symbols as "children";

(ii) "twin" the two least probable children and give the twin the sum of the probabilities:

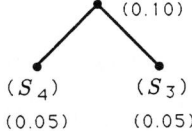

(iii) regard the twin as a child; and

(iv) repeat steps (ii) and (iii) until all children are accounted for.

This tree is now labeled with 1's and 0's to obtain the Huffman code. The labeling procedure is to label each right branch with a 1 and each left branch with a 0. The procedure for laying out symbols and constructing Huffman trees and codes is illustrated in the following examples.

Example 7.2. Consider the source statistics

Symbol	S_0	S_1	S_2	S_3	S_4
Probability	0.5	0.2	0.2	0.05	0.05

for which the Huffman algorithm produces the following binary tree and its corresponding code:

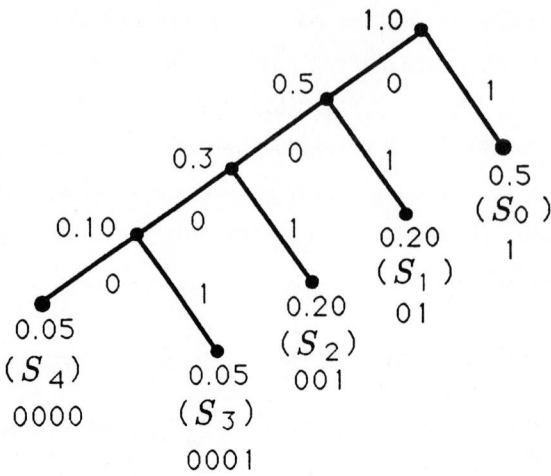

Example 7.3. The Huffman code for the source statistics

Symbol	S_0	S_1	S_2	S_3	S_4
Probability	0.75	0.075	0.075	0.05	0.05

is illustrated next:

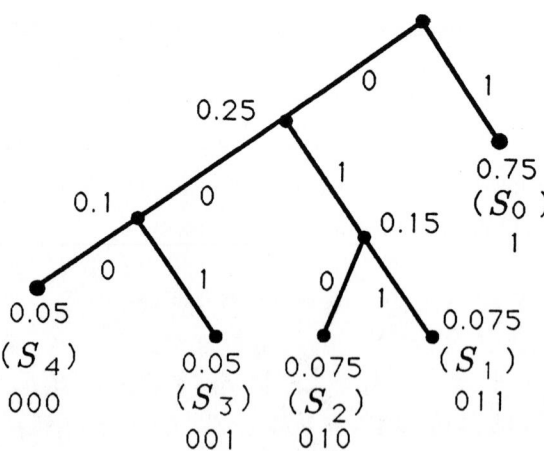

Problem 7.11   Generate binary trees and Huffman codes for the following source statistics:

Symbol	S_0	S_1	S_2	S_3	S_4	S_5	S_6	S_7
Probability 1	0.20	0.20	0.15	0.15	0.1	0.1	0.05	0.05
Probability 2	0.3	0.25	0.1	0.1	0.075	0.075	0.05	0.05 . ∎

Coding a FAX Machine. Symbols can arise in unusual ways and be defined quite arbitrarily. To illustrate this point, we consider the design of a hypothetical FAX machine. For our design we will assume that a laser scanner reads a page of black and white text or pictures, producing a high voltage for a black spot and a low voltage for a white spot. We will also assume that the laser scanner resolves the page at 1024 lines, with 1024 spots/line. This means that each page is represented by a two-dimensional array, or matrix, of pixels (picture elements), each pixel being 1 or 0. If we simply transmitted these 1's and 0's, then we would need $1024 \times 1024 = 1,059,576$ bits. If these were transmitted over a 9600 baud phone line, then it would take almost 2 minutes to transmit the FAX. This is a long time.

Let's think about a typical scan line for a printed page. It will contain long runs of 0's, corresponding to long runs of white, interrupted by short bursts of 1's, corresponding to short runs of black where the scanner encounters a line or a part of a letter. So why not try to define a symbol to be "a run of k 0's" and code these runs? The resulting code is called a "run length code." Let's define eight symbols, corresponding to run lengths from 0 to 7 (a run length of 0 is a 1):

$$S_0 = \text{run length of 0 zeros (a 1)}$$
$$S_1 = \text{run length of 1 zero}$$
$$\vdots$$
$$S_7 = \text{run length of 7 zeros.}$$

7.14

If we simply used a simple three-bit binary code for these eight symbols, then for each scan line we would generate anywhere from 3×1024 bits (for a scan line consisting of all 1's) to $3 \times 1024/7 \cong 400$ bits (for a scan line consisting of all 0's). But what if we ran an experiment to determine the relative frequency of the run lengths S_0 through S_7 and used a Huffman code to "run length encode" the run lengths? The following problem explores this possibility and produces an efficient FAX code.

Problem 7.12 An experiment conducted on FAXed documents produces the following statistics for run lengths of white ranging from 0 to 7:

Symbol	S_0	S_1	S_2	S_3	S_4	S_5	S_6	S_7
Probability	0.01	0.06	0.1	0.1	0.2	0.15	0.15	0.2

These statistics indicate that only 1% of a typical page is black. Construct the Huffman code for this source. Use your Huffman code to code and decode these scan lines:

7.5 Hamming Codes for Channel Coding

The idea behind Hamming codes is to intersperse, or append, extra binary digits to a binary code so that errors in transmission of the code over a channel may be detected and corrected. For example, suppose we transmit the code 01101001, and it is received as 01001001. In this transmission, the third most significant bit is received erroneously. Let's define the following "modulo-2 addition" of binary numbers:

$$0 \oplus 0 = 0 \qquad (7.15)$$
$$0 \oplus 1 = 1$$
$$1 \oplus 0 = 1$$
$$1 \oplus 1 = 0.$$

Multiplication in modulo-2 arithmetic is simply $0 \cdot 0 = 0 \cdot 1 = 1 \cdot 0 = 0$ and $1 \cdot 1 = 1$. Then we can say that the *error sequence* 00100000 is "added" to the *transmission* 01101001 to produce the erroneous reception:

$$
\begin{array}{ll}
\ 01101001 & \text{transmitted} \\
\oplus\ 00100000 & \text{error} \\
\hline
\ 01001001 & \text{received.}
\end{array}
$$

7.16

Hamming error correcting codes will permit us to receive the erroneous transmission and to detect and correct the error. This is obviously of great value in transmitting and storing information. (Imagine how upset you would be to have the binary code for your checking account confused with that of Mrs. Joan Kroc.)

Choosing the Number of Check Bits. Let's suppose we have N bits of information that we wish to transmit and that we wish to intersperse "check bits" that will enable us to detect and correct any single bit error in the transmission. If we use N information bits and n check bits, then we will transmit a code word containing $N + n$ bits. The n check bits can code 2^n events, and we want these events to indicate whether or not any errors occurred and, if so, where they occurred. Therefore we require

$$2^n \geq (N + n) + 1 \qquad\qquad 7.17$$

where $(N + n)$ is the number of single error events that can occur and $+1$ is the number of no-error events. For example, when $N = 4$, we require $n = 3$ so that $2^3 \geq (4 + 3) + 1$.

Problem 7.13 How many check bits do you require to code seven bits of information for single error correction? ∎

Code Construction. Let's suppose we have constructed an (N, n) Hamming code consisting of N information bits and n check bits (or parity bits). We denote the information bits by $x_1, x_2, \ldots, x_N$ and the check bits by $c_1, c_2, \ldots, c_n$. These bits may be interspersed. When $N = 4$ and $n = 3$, then a typical array of bits within a code word would be one of the following:

$$
\begin{bmatrix} c_1 \\ c_2 \\ x_1 \\ c_3 \\ x_2 \\ x_3 \\ x_4 \end{bmatrix} \quad \text{or} \quad \begin{bmatrix} x_1 \\ x_2 \\ x_3 \\ x_4 \\ c_1 \\ c_2 \\ c_3 \end{bmatrix}. \qquad 7.18
$$

The first ordering is "natural" (as we will see), and the second is "systematic" (a term that is used to describe any code whose head is information and whose tail is check). If a single error occurs in an (N, n) code, then the received code word will be the modulo-2 sum of the code word and the error word that contains a 1 in its ith position:

$$
\begin{bmatrix} c_1 \\ c_2 \\ x_1 \\ c_3 \\ x_2 \\ x_3 \\ x_4 \end{bmatrix} \oplus \begin{bmatrix} 0 \\ 0 \\ 0 \\ 1 \\ 0 \\ 0 \\ 0 \end{bmatrix}. \qquad 7.19
$$

We would like to operate on this received code word in such a way that the location of the error bit can be determined. If there were no code word, then an obvious solution would be to premultiply the error word by the *parity check* matrix

$$
\mathbf{A}^T = \begin{bmatrix} 1 & 0 & 1 & 0 & 1 & 0 & 1 \\ 0 & 1 & 1 & 0 & 0 & 1 & 1 \\ 0 & 0 & 0 & 1 & 1 & 1 & 1 \end{bmatrix}. \qquad 7.20
$$
$$
\phantom{\mathbf{A}^T = } (1) \quad (2) \quad (3) \quad (4) \quad (5) \quad (6) \quad (7)
$$

The ith column of $\mathbf{A}^T$ is just the binary code for i. When $\mathbf{A}^T$ premultiplies an error word, the error bit picks out the column that codes the error position:

$$\begin{bmatrix} 1 & 0 & 1 & 0 & 1 & 0 & 1 \\ 0 & 1 & 1 & 0 & 0 & 1 & 1 \\ 0 & 0 & 0 & 1 & 1 & 1 & 1 \end{bmatrix} \begin{bmatrix} 0 \\ 0 \\ 0 \\ 1 \\ 0 \\ 0 \\ 0 \end{bmatrix} = \begin{bmatrix} 0 \\ 0 \\ 1 \end{bmatrix} \qquad 7.21$$

position 4 → , binary code for position 4

If the error word contains no error bits, then the product is 0, indicating no errors.

This seems like a good idea, but what about the effect of the code word? In Problem 7.14, you are asked to show that the effect of the parity check matrix $\mathbf{A}^T$ applied to the modulo-2 sum of a code word $\mathbf{x}$ and an error word $\mathbf{e}$ is

$$\mathbf{A}^T(\mathbf{x} \oplus \mathbf{e}) = \mathbf{A}^T\mathbf{x} \oplus \mathbf{A}^T\mathbf{e}. \qquad 7.22$$

In this equation all sums and products obey the rules of modulo-2 arithmetic.

Problem 7.14 Let $\mathbf{y} = \mathbf{x} \oplus \mathbf{e}$ denote the modulo-2 sum of a code word $\mathbf{x}$ and an error word $\mathbf{e}$; $\mathbf{A}^T$ is a parity check matrix. Show that

$$\mathbf{A}^T\mathbf{y} = \mathbf{A}^T\mathbf{x} \oplus \mathbf{A}^T\mathbf{e}. \quad \blacksquare$$

We have designed the parity check matrix $\mathbf{A}^T$ so that the *syndrome* $\mathbf{A}^T\mathbf{e}$ produces a binary code for the error location. (The location of the error is the syndrome for the error word.) The product $\mathbf{A}^T\mathbf{x}$ will interfere with this syndrome unless $\mathbf{A}^T\mathbf{x} = \mathbf{0}$. Therefore we will require that the code word $\mathbf{x}$ satisfy the constraint

$$\mathbf{A}^T\mathbf{x} = \mathbf{0}. \qquad\qquad 7.23$$

This constraint actually *defines* the Hamming code. Let's illustrate this point by applying the constraint to a code word in its "natural format" $\mathbf{x}^T = (c_1c_2x_1c_3x_2x_3x_4)$.

Natural Codes. When the information bits and the check bits are coded in their natural order $(c_1c_2x_1c_3x_2x_3x_4)$, then we may determine the check bits by writing $\mathbf{A}^T\mathbf{x}$ as follows:

$$\begin{bmatrix} 1 & 0 & 1 & 0 & 1 & 0 & 1 \\ 0 & 1 & 1 & 0 & 0 & 1 & 1 \\ 0 & 0 & 0 & 1 & 1 & 1 & 1 \end{bmatrix} \begin{bmatrix} c_1 \\ c_2 \\ x_1 \\ c_3 \\ x_2 \\ x_3 \\ x_4 \end{bmatrix} = \begin{bmatrix} 0 \\ 0 \\ 0 \end{bmatrix}. \qquad 7.24$$

We use the rules of modulo-2 arithmetic to write these constraints as

$$c_1 \oplus x_1 \oplus x_2 \oplus x_4 = 0 \qquad\qquad 7.25$$
$$c_2 \oplus x_1 \oplus x_3 \oplus x_4 = 0$$
$$c_3 \oplus x_2 \oplus x_3 \oplus x_4 = 0.$$

Therefore the check bits c_1, c_2, and c_3 are simply the following modulo-2 sums:

$$c_1 = x_1 \oplus x_2 \oplus x_4 \qquad\qquad 7.26$$
$$c_2 = x_1 \oplus x_3 \oplus x_4$$
$$c_3 = x_2 \oplus x_3 \oplus x_4.$$

This finding may be organized into the matrix equation

$$
\begin{bmatrix} c_1 \\ c_2 \\ x_1 \\ c_3 \\ x_2 \\ x_3 \\ x_4 \end{bmatrix} = \begin{bmatrix} 1 & 1 & 0 & 1 \\ 1 & 0 & 1 & 1 \\ 1 & 0 & 0 & 0 \\ 0 & 1 & 1 & 1 \\ 0 & 1 & 0 & 0 \\ 0 & 0 & 1 & 0 \\ 0 & 0 & 0 & 1 \end{bmatrix} \begin{bmatrix} x_1 \\ x_2 \\ x_3 \\ x_4 \end{bmatrix}. \qquad 7.27
$$

This equation shows how the code word $\mathbf{x}$ is built from the information bits (x_1, x_2, x_3, x_4). We call the matrix that defines the construction a *coder matrix* and write it as $\mathbf{H}$:

$$
\mathbf{x} = \mathbf{H}\boldsymbol{\Theta} \qquad 7.28
$$

$$
\mathbf{x}^T = (c_1 \; c_2 \; x_1 \; c_3 \; x_2 \; x_3 \; x_4) \qquad \boldsymbol{\Theta}^T = (x_1 \; x_2 \; x_3 \; x_4)
$$

$$
\mathbf{H} = \begin{bmatrix} 1 & 1 & 0 & 1 \\ 1 & 0 & 1 & 1 \\ 1 & 0 & 0 & 0 \\ 0 & 1 & 1 & 1 \\ 0 & 1 & 0 & 0 \\ 0 & 0 & 1 & 0 \\ 0 & 0 & 0 & 1 \end{bmatrix}. \qquad 7.29
$$

This summarizes the construction of a Hamming code $\mathbf{x}$.

Problem 7.15 Check to see that the product of the parity check matrix $\mathbf{A}^T$ and the coder matrix $\mathbf{H}$ is $\mathbf{A}^T\mathbf{H} = \mathbf{0}$. Interpret this result. ∎

Problem 7.16 Fill in the following table to show what the Hamming $(4, 3)$ code is:

x_1	x_2	x_3	x_4	c_1	c_2	x_1	c_3	x_2	x_3	x_4
0	0	0	0	0	0	0	0	0	0	0
0	0	0	1	1	1	0	1	0	0	1
0	0	1	0							
0	0	1	1							
0	1	0	0							
0	1	0	1							
0	1	1	0							
0	1	1	1							
1	0	0	0							
1	0	0	1							
1	0	1	0							
1	0	1	1							
1	1	0	0							
1	1	0	1							
1	1	1	0							
1	1	1	1							∎

Problem 7.17 Design a Hamming $(11, n)$ code for coding eleven information bits against single errors. Show your equations for $c_1, c_2, \ldots, c_n$ and write out the coder matrix $\mathbf{H}$ for $\mathbf{x} = \mathbf{H\Theta}$. ∎

Decoding. To decode a Hamming code, we form the syndrome $\mathbf{A}^T\mathbf{y}$ for the received (and possibly erroneous) code word $\mathbf{y} = \mathbf{x} \oplus \mathbf{e}$. Because $\mathbf{A}^T\mathbf{x} = \mathbf{0}$, the syndrome is

$$\mathbf{s} = \mathbf{A}^T\mathbf{e}. \qquad\qquad 7.30$$

Convert this binary number into its corresponding integer location and change the bit of $\mathbf{y}$ in that location. If the location is zero, do nothing. Now strip off the information bits. This is the decoding algorithm.

Problem 7.18   Use the table of Hamming (4, 3) codes from Problem 7.16 to construct a table of *received* codes that contain either no bit errors or exactly one bit error. Apply the decoding algorithm to construct (x_1, x_2, x_3, x_4) and show that all received code words with one or fewer errors are correctly decoded. ∎

Digital Hardware. The tables you have constructed in Problems 7.16 and 7.18 for coding and decoding Hamming $(4, 3)$ codes may be stored in digital logic chips. Their functionality is illustrated in Figure 7.9. The coder chip accepts $(x_1x_2x_3x_4)$ as its address and generates a coded word. The decoder chip accepts $(c_1c_2x_1c_3x_2x_3x_4)$ as its address and generates a decoded word. In your courses on digital logic you will study circuits for implementing coders and decoders.

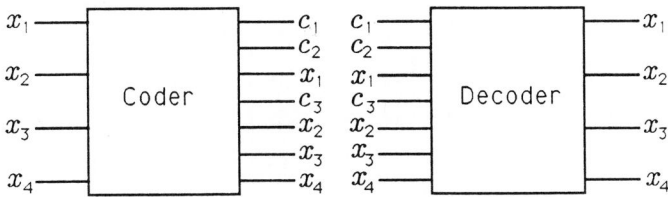

Figure 7.9: Digital Logic for Hamming Code

Problem 7.19 Discuss the possibility of detecting a received $(4, 3)$ code word that is neither a valid code word nor a code word with a single error. How would you use such a detector? ∎

Problem 7.20   What fraction of received seven-bit words can be correctly decoded as Hamming $(4, 3)$ codes? ∎

Systematic Codes. Systematic Hamming codes are codes whose information bits lead and whose check bits trail. The format for a $(4, 3)$ code is then $(x_1 x_2 x_3 x_4 c_1 c_2 c_3)$. The construction of a $(4, 3)$ code word from the information bits may be written as

$$
\begin{bmatrix} x_1 \\ x_2 \\ x_3 \\ x_4 \\ c_1 \\ c_2 \\ c_3 \end{bmatrix} = \begin{bmatrix} 1 & 0 & 0 & 0 \\ 0 & 1 & 0 & 0 \\ 0 & 0 & 1 & 0 \\ 0 & 0 & 0 & 1 \\ c_{11} & c_{12} & c_{13} & c_{14} \\ c_{21} & c_{22} & c_{23} & c_{24} \\ c_{31} & c_{32} & c_{33} & c_{34} \end{bmatrix} \begin{bmatrix} x_1 \\ x_2 \\ x_3 \\ x_4 \end{bmatrix}.
\tag{7.31}
$$

The coder matrix takes the form

$$
\mathbf{H} = \begin{bmatrix} 1 & 0 & 0 & 0 \\ 0 & 1 & 0 & 0 \\ 0 & 0 & 1 & 0 \\ 0 & 0 & 0 & 1 \\ c_{11} & c_{12} & c_{13} & c_{14} \\ c_{21} & c_{22} & c_{23} & c_{24} \\ c_{31} & c_{32} & c_{33} & c_{34} \end{bmatrix} = \begin{bmatrix} \mathbf{I} \\ - \\ \mathbf{C} \end{bmatrix}.
\tag{7.32}
$$

The problem is to find the matrix $\mathbf{C}$ that defines the construction of check bits. The constraint $\mathbf{A}^T \mathbf{x} = \mathbf{0}$ produces the constraint $\mathbf{A}^T \mathbf{H} = \mathbf{0}$ so that $\mathbf{A}^T \mathbf{H} \boldsymbol{\Theta} = \mathbf{0}$. The constraints $\mathbf{A}^T \mathbf{H} = \mathbf{0}$ may be written out as

$$
\begin{bmatrix} 1 & 0 & 1 & 0 & 1 & 0 & 1 \\ 0 & 1 & 1 & 0 & 0 & 1 & 1 \\ 0 & 0 & 0 & 1 & 1 & 1 & 1 \end{bmatrix} \begin{bmatrix} 1 & 0 & 0 & 0 \\ 0 & 1 & 0 & 0 \\ 0 & 0 & 1 & 0 \\ 0 & 0 & 0 & 1 \\ c_{11} & c_{12} & c_{13} & c_{14} \\ c_{21} & c_{22} & c_{23} & c_{24} \\ c_{31} & c_{32} & c_{33} & c_{34} \end{bmatrix} = \begin{bmatrix} 0 & 0 & 0 & 0 \\ 0 & 0 & 0 & 0 \\ 0 & 0 & 0 & 0 \end{bmatrix}.
\tag{7.33}
$$

These constraints produce all the equations we need (twelve equations in twelve unknowns) to determine the c_{ij}.

Problem 7.21 Solve Equation 7.33 for the c_{ij}. Show that the coder matrix for a systematic Hamming $(4,3)$ code is

$$\mathbf{H} = \begin{bmatrix} 1 & 0 & 0 & 0 \\ 0 & 1 & 0 & 0 \\ 0 & 0 & 1 & 0 \\ 0 & 0 & 0 & 1 \\ 0 & 1 & 1 & 1 \\ 1 & 0 & 1 & 1 \\ 1 & 1 & 0 & 1 \end{bmatrix} . \quad \blacksquare$$

Problem 7.22 Show that the coder matrix of Problem 7.20 is a permutation of the coder matrix in Equation 7.29. (That is, the rows are reordered.) $\blacksquare$

Problem 7.23 (MATLAB) Write a MATLAB program that builds Hamming $(4,3)$ codes from information bits $(x_1x_2x_3x_4)$ and decodes Hamming $(4,3)$ codes $(c_1c_2x_1c_3x_2x_3x_4)$ to obtain information bits $(x_1x_2x_3x_4)$. Synthesize all seven-bit binary codes and show that your decoder correctly decodes correct codes and one-bit error codes. $\blacksquare$

7.6 Numerical Experiment (Huffman Codes)

Table 7.3 contains Lincoln's Gettysburg Address. Ignore special symbols like periods and ignore the distinction between lowercase and uppercase letters. Compute the relative frequency of occurrence for each of the 26 letters of the alphabet. Write a MATLAB program to generate a binary tree for the Huffman code of the Gettysburg Address. When you have generated the code, compute the average code word length

$$L = \sum_{i=1}^{26} \frac{n_i}{N} l_i$$

where $\frac{n_i}{N}$ is the relative frequency for symbol i and l_i is the code word length for symbol i. Compare L with the entropy

$$H = -\sum_{i=1}^{26} \frac{n_i}{N} \log_2 \frac{n_i}{N} = \sum_{i=1}^{26} \frac{n_i}{N} \log_2 \frac{N}{n_i}$$

and compare L to $\log_2 26$. Interpret your findings.

Table 7.3: Gettysburg Address

Fourscore and seven years ago, our fathers brought forth upon this continent a new nation, conceived in liberty and dedicated to the proposition that all men are created equal. Now we are engaged in a great civil war, testing whether that nation or any nation so conceived and so dedicated can long endure. We are met on a great battlefield of that war. We have come to dedicate a portion of that field as a final resting-place for those who have given their lives that that nation might live. It is altogether fitting and proper that we should do this. But in a larger sense, we cannot dedicate, we cannot consecrate, we cannot hallow this ground. The brave men, living and dead, who struggled here have consecrated it far above our power to add or detract. The world will little note nor long remember what we say here, but it can never forget what they did here. It is for us the living rather to be dedicated here to the unfinished work which they who fought here have thus far so nobly advanced. It is rather for us to be here dedicated to the great task remaining before us—that from these honored dead we take increased devotion to that cause for which they gave the last full measure of devotion—that we here highly resolve that these dead should not have died in vain, that this nation under God shall have a new birth of freedom, and that government of the people, by the people, for the people shall not perish from the Earth.

–Abraham Lincoln, November 19, 1863

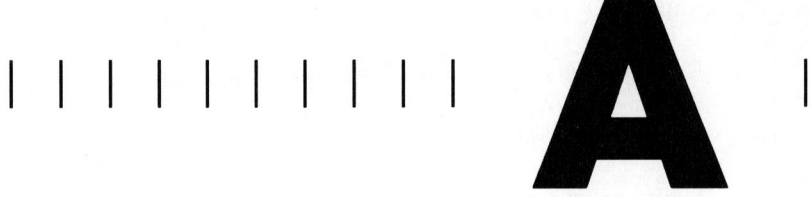

An Introduction to MATLAB

A.1 Introduction

MATLAB stands for "Matrix Laboratory." It is a computing environment specifically designed for matrix computations. The program is ideally suited to circuit analysis, signal processing, filter design, control system analysis, and much more. Beyond that, its versatility with complex numbers and graphics makes it an attractive choice for many other programming tasks. MATLAB can be thought of as a programming language like PASCAL, FORTRAN, C, or BASIC. Like most versions of BASIC, MATLAB can be used in an interactive mode wherein statements are executed immediately as they are typed. Alternatively, a program can be written in advance and saved to a disc file using an editor and then executed in MATLAB. You will find both modes of operation useful.

A.2 Running MATLAB (Macintosh)

In order to run MATLAB on a Macintosh SE or PLUS computer, you need the program called EDU-MATLAB. The program requires at least 1 Mbyte of memory, System 3.0 or above, Finder version 3.0 or above, and an 800K drive. A hard disc drive is highly recommended. In order to run

Acknowledgment: This appendix, written with assistance from Cédric J. Demeure and Peter Massey, was inspired by the MATLAB user's manual from The MATHWORKS, Inc. The *MATLAB Primer*, available through the MATLAB User's Group, is a useful learning aid for teachers and students. To join the MATLAB User's Group, send your request via E-mail to matlab_users_request@mcs.anl.gov.

MATLAB on a Macintosh II, IIx, IIcx, or SE/30, you need the program called MacII-MATLAB, and the same system requirements apply.

To start MATLAB, you may need to open the folder containing the MATLAB program. Then just "double-click" the program icon or the program name (for example, EDU-MATLAB). Figure A.1 shows a typical organization of the folder containing Mac II-MATLAB. It contains the main program, the settings file, the demonstrations folder, and any toolbox folders. The double-click on Mac II-MATLAB produces the *Command* window as shown in Figure A.2. You will also see a *Graph* window partially hidden behind it. (The fact that the window is not in front means that it is opened but not currently active.) If you do not know what "clicking," "dragging," "popup menu," and "trash" mean, you should stop reading now and familiarize yourself with the Macintosh.

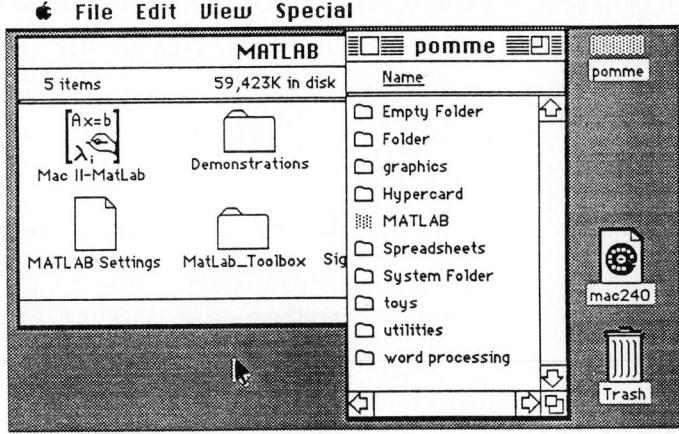

Figure A.1: The MATLAB Folder (©Apple Computer, Inc., used with permission.)

In the command window, you should see the prompt "≫." The program interpreter is waiting for you to enter instructions. At this point it is a good idea to run the demonstration programs that are available in the "About MATLAB" menu under the Apple menu. Just click on the "demos" button and select a demo. During pauses, strike any key to continue. Whenever you have a MATLAB file in any folder, then you may double-click the file to launch the program. This allows you to have your own folder containing your own MATLAB files, separated from the MATLAB folder.

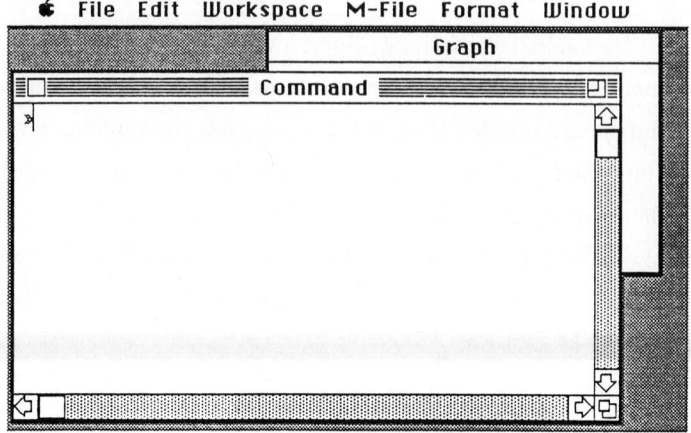

Figure A.2: The Command Window (©Apple Computer, Inc., used with permission.)

MATLAB has four types of windows:

(i) *Command* for computing, programming, and designing input/output displays;

(ii) *Graph* for displaying plots and graphs;

(iii) *Edit* for creating and modifying your own files; and

(iv) *Help* for getting on-line help and for running demos.

All windows follow the traditional behavior of Macintosh windows. You can resize them (actually the help window has a fixed size) or move them. For more details on menus and windows, see the Macintosh and MATLAB manuals.

A.3 Running MATLAB (PC)

In order to run MATLAB (Version 3.5) on an IBM or compatible personal computer, you must have a floating point math coprocessor (80x87) installed and at least 512 kbytes of memory. The program is called PCMAT-LAB.EXE, but it is usually invoked via the batch file MATLAB.BAT in the MATLAB subdirectory. If you are using a menu system and MATLAB is one of your choices, just choose it. Otherwise, go to the MATLAB subdirectory and type MATLAB.

You may be able to use a more powerful implementation of MATLAB if you have an 80286 or 80386 machine. AT-MATLAB runs on an 80286 with at least 1 Mbyte of extended memory. AT-MATLAB is distributed with PC-MATLAB. 386-MATLAB, a special version for 80386 or 80486 machines with virtual memory support and no limits on variable size, is sold separately.

When you run MATLAB, you should see the prompt "≫." The program interpreter is waiting for you to enter instructions. Some MATLAB instructions, such as `plot`, are graphics-type instructions which plot results and data. Execution of one of these graphics instructions puts the PC screen into the *graphics mode*, which displays the resulting plot. No instructions can be executed in the graphics mode other than a screen-dump function. Striking any other key will return the PC to the *command mode*, but the graphics are temporarily stored (like variables) and can be recalled by the `shg` (show graphics) instruction. If you wish, you may run some of the demonstration programs now by entering `demo` and following the on-screen instructions.

A.4 Interactive Mode

In command mode, MATLAB displays a prompt (≫) and waits for your input. You may type any legal mathematical expression for immediate evaluation. Try the following three examples (press "enter" or "return" at the end of each line):

```
≫ 2+2
≫ 5^2
≫ 2*sin(pi/4)
```

The variable `pi` = `3.14`... is built into MATLAB, as are the `sin` function and hundreds of other functions. When you entered each of the preceding lines,

MATLAB stored the results in a variable called **ans** for *answer*. The value of **ans** was then displayed. The last line should have produced the square root of 2. We can manipulate **ans** to find out

>> ans^2

The new answer is very close to 2, as expected. Let's see what the roundoff error is:

>> ans-2

A.5 Variables

Any result you wish to keep for a while may be assigned to a variable other than **ans**:

```
>> x = pi/7
>> cos(x)
>> y = sin(x)^2+cos(x)^2;
>> y
```

A semicolon (;) at the end of the line suppresses printing of the result, as when we calculated **y** in the next-to-last line just shown. This feature is especially useful when writing MATLAB programs where intermediate results are not of interest and when working with large matrices.

MATLAB supports the dynamic creation of variables. You can create your own variables by just assigning a value to a variable. For example, type **x = 3.5+4.2**. Then the real variable **x** contains the value **7.7**. Variable names must start with an alphabetical character and be less than nineteen characters long. If you type **x = -3*4.0**, the content **7.7** is replaced by the value **-12**. Some commands allow you to keep track of all the variables that you have already created in your session. Type **who** or **whos** to get the list and names of the variables currently in memory (**whos** gives more information than **who**). To clear all the variables, type in **clear**. To clear a single variable (or several) from the list, follow the command **clear** by the name of the variable you want to delete or by a list of variable names separated by spaces. Try it now.

MATLAB is case sensitive. In other words, x and X are two different variables. You can control the case sensitivity of MATLAB by entering the command `casesen`, which toggles the sensitivity. The command `casesen on` enforces case sensitivity, and `casesen off` cancels it.

If one line is not enough to enter your command, then finish the first line with two dots (..) and continue on the next line. You can enter more than one command per line by separating them with commas if you want the result displayed or with semicolons if you do not want the result displayed. For example, type

```
≫ theta = pi/7; x = cos(theta); y = sin(theta);
≫ x,y
```

to first compute `theta`, `cos(theta)`, and `sin(theta)` and then to print x and y.

A.6 Complex Variables

The number $\sqrt{-1}$ is predefined in MATLAB and stored in the two variable locations denoted by i and j. This double definition comes from the preference of mathematicians for using i and the preference of engineers for using j (with i denoting electrical current). i and j are variables, and their contents may be changed. If you type j = 5, then this is the value for j and j no longer contains $\sqrt{-1}$. Type in j = `sqrt(-1)` to restore the original value. Note the way a complex variable is displayed. If you type i, you should get the answer

```
i =
    0+1.0000i.
```

The same value will be displayed for j. Try it. Using j, you can now enter complex variables. For example, enter z1 = 1+2*j and z2 = 2+1.5*j. As j is a variable, you have to use the multiplication sign *. Otherwise, you will get an error message. MATLAB does not differentiate (except in storage) between a real and a complex variable. Therefore variables may be added, subtracted, multiplied, or even divided. For example, type in x = 2, z = 4.5*j, and z/x. The real and imaginary parts of z are both divided by x. MATLAB just treats the real variable x as a complex variable with a zero imaginary part. A complex variable that happens to have a zero imaginary part is treated like a real variable. Subtract 2*j from z1 and display the result.

MATLAB contains several built-in functions to manipulate complex numbers. For example, real(z) extracts the real part of the complex number z. Type

```
≫ z = 2+1.5*j, real(z)
```

to get the result

```
z =
   2.000+1.500i

ans =
   2
```

Similarly, imag(z) extracts the imaginary part of the complex number z. The functions abs(z) and angle(z) compute the absolute value (magnitude) of the complex number z and its angle (in radians). For example, type

```
≫ z = 2+2*j;
≫ r = abs(z)
≫ theta = angle(z)
≫ z = r*exp(j*theta)
```

The last command shows how to get back the original complex number from its magnitude and angle. This is clarified in Chapter 1: Complex Numbers.

Another useful function, conj(z), returns the complex conjugate of the complex number z. If z = x+j*y where x and y are real, then conj(z) is equal to x-j*y. Verify this for several complex numbers by using the function conj(z).

A.7 Vectors and Matrices

As its name indicates, MATLAB is especially designed to handle matrices. The simplest way to enter a matrix is to use an explicit list. In the list, the elements are separated by blanks or commas, and the semicolon (;) is used to indicate the end of a row. The list is surrounded by square brackets [ ]. For example, the statement

    ≫ A = [1 2 3;4 5 6;7 8 9]
results in the output

    A =
       1 2 3
       4 5 6
       7 8 9

The variable A is a matrix of size 3×3. Matrix elements can be any MATLAB expression. For example, the command

    ≫ x = [-1.3 sqrt(3) (1+2+3)*4/5]
results in the matrix

    x =
       -1.3000 1.7321 4.8000

We call a matrix with just one row or one column a vector, and a 1×1 matrix is a scalar. Individual matrix elements can be referenced with indices that are placed inside parentheses. Type x(5) = abs(x(1)) to produce the new vector

    x =
       -1.3000 1.7321 4.8000 0.000 1.3000

Note that the size of x has been automatically adjusted to accommodate the new element and that elements not referenced are set equal to 0 (here x(4)). New rows or columns can be added very easily. Try typing r = [10 11 12],A = [A;r]. Dimensions in the command must coincide. Try r = [13 14],A = [A;r].

The command `size(A)` gives the number of rows and the number of columns of A. The output from `size(A)` is itself a matrix of size 1×2. These numbers can be stored if necessary by the command `[m n] = size(A)`. In our previous example, `A = [A;r]` is a 4×3 matrix, so the variable m will contain the number 4 and n will contain the number 3. A vector is a matrix for which either m or n is equal to 1. If m is equal to 1, the matrix is a row vector; if n is equal to 1, the matrix is a column vector. Matrices and vectors may contain complex numbers. For example, the statement

>> A = [1 2;3 4]+j*[5 6;7 8]

and the statement

>> A = [1+5*j 2+6*j;3+7*j 4+8*j]

are equivalent, and they both produce the matrix

```
A =
   1.0000+5.0000i    2.0000+6.0000i
   3.0000+7.0000i    4.0000+8.0000i
```

Note that blanks must be avoided in the second expression for A. Try typing

>> A = [1 + 5*j 2 + 6*j;3 +7*j 4 + 8*j]

What is the size of A now?

MATLAB has several built-in functions to manipulate matrices. The special character, ', for prime denotes the transpose of a matrix. The statement `A = [1 2 3;4 5 6;7 8 9]'` produces the matrix

```
A =
   1 4 7
   2 5 8
   3 6 9
```

The rows of A' are the column of A, and vice versa. If A is a complex matrix, then A' is its complex conjugate transpose or hermitian transpose. For an "unconjugate" transpose, use the two-character operator dot-prime (.'). Matrix and vector variables can be added, subtracted, and multiplied as regular variables if the sizes match. Only matrices of the same size can be added or subtracted. There is, however, an easy way to add or subtract a common scalar from each element of a matrix. For example, x = [1 2 3 4],x = x-1 produces the output

```
x =
   1 2 3 4

x =
   0 1 2 3
```

As discussed in Chapter 4, multiplication of two matrices is only valid if the inner sizes of the matrices are equal. In other words, A*B is valid if the second size of A (number of columns) is the same as the first size of B (number of rows). Let $a_{i,j}$ represent the element of A in the ith row and the jth column. Then the matrix A*B consists of elements

$$(AB)_{i,j} = \sum_{k=1}^{n} a_{i,k} b_{k,j}$$

where n is the number of columns of A and the number of rows of B. Try typing A = [1 2 3;4 5 6]; B = [7;8;9]; A*B. You should get the result

```
ans =
    50
   112
```

The inner product between two column vectors x and y is the scalar defined as the product x'*y or equivalently as y'*x. For example, x = [1;2],y = [3;4],x'*y, leads to the result

```
ans =
    11
```

Similarly, for row vectors the inner product is defined as x*y'. The Euclidean norm of a vector is defined as the square root of the inner product between a vector and itself. Try to compute the norm of the vector [1 2 3 4]. You should get 5.4772. The outer product of two column (row) vectors is the matrix x*y' (x'*y).

Any scalar can multiply or be multiplied by a matrix. The multiplication is then performed element by element. Try A = [1 2 3;4 5 6;7 8 9];A*2. You should get

```
ans =
    2    4    6
    8   10   12
   14   16   18
```

Verify that 2*A gives the same result.

The inverse of a matrix is computed by using the function inv(A) and is only valid if A is square. If the matrix is singular, meaning that it has no inverse, a message will appear. Try typing inv(A). You should get

```
Warning: Matrix is close to singular or badly scaled.
    Results may be inaccurate. RCOND=2.937385e-18

ans =
    1.0e+16*
    0.3152   -0.6304    0.3152
   -0.6304    1.2609   -0.6304
    0.3152   -0.6304    0.3152
```

The inverse of a matrix may be used to solve a linear system of equations. For example, to solve the system

$$\begin{bmatrix} 1 & 2 & 3 \\ 1 & -2 & 4 \\ 0 & -2 & 1 \end{bmatrix} \begin{bmatrix} x_1 \\ x_2 \\ x_3 \end{bmatrix} = \begin{bmatrix} 2 \\ 7 \\ 3 \end{bmatrix},$$

you could type A = [1 2 3;1 -2 4;0 -2 1]; b = [2;7;3]; inv(A)*b and get

```
ans =
    1
   -1
    1
```

Check to see that this is the correct answer by typing A*[1;-1;1]. What do you see?

MATLAB offers another way to solve linear systems, based on Gauss elimination, that is faster than computing the inverse. The syntax is `A\b` and is valid whenever `A` has the same number of rows as `b`. Try it.

The "Dot" Operator. Sometimes you may want to perform an operation element by element. In MATLAB, these element-by-element operations are called array operations. Of course, matrix addition and subtraction are already element-by-element operations. The operation `A.*B` denotes the multiplication, element by element, of the matrices `A` and `B`. Make two 3×3 matrices, **A** and **B**, and try

```
>> A*B
>> A.*B
```

Suppose we want to find the square of each number in **A**. The proper way to specify this calculation is

```
>> A_squared=A.^2
```

where the period (dot) indicates an "array operation" to be performed on each element of `A`. Without the dot, `A` is multiplied by `A` according to the rules of matrix multiplication described in Chapter 4, giving a totally different result:

```
>> A^2
```

Subtleties. Because MATLAB can do so many different mathematical functions with just a few keystrokes, there are times when a very slight change in what you type will lead to a different result. Using the matrix `A` entered earlier, type the following two lines:

```
>> 2.^A
>> 2 .^A                %with a space after the 2.
```

In the first case, the dot is "absorbed" by the 2 as a decimal point and the ^ is taken as a matrix exponential. But, when the dot is separated from the 2 by a space, it becomes part of the operator (`.^`) and specifies that 2 should be raised to the power of each element in **A**. The point is, you should be very careful to type what you mean in an unambiguous way until you are familiar enough with MATLAB to know how the subtle situations will be interpreted. An unambiguous way of typing the preceding lines is

```
>> (2.)^A          %for matrix exponential
>> (2).^A          %for array exponential.
```

A.8 The Colon

You can use the colon several ways in MATLAB (see `help :`). Its basic meaning is a vector of sequential values. For example, type

```
>> x = 3:9
```

to get

```
x =
   3  4  5  6  7  8  9
```

For increments other than 1, use statements like

```
>> x = 1:0.5:4
>> x = 6:-1:0
```

Most MATLAB functions will accept vector inputs and produce vector outputs. The statement

```
>> y = sqrt(1:10)
```

builds a vector of integers from 1 to 10 and takes the square root of each of those numbers. Try it.

Now for another subtlety—what is the effect of each of the following statements and why?

```
>> 1+1:5
>> 1+(1:5)
```

Appending to a Matrix or Vector. A matrix or vector can be enlarged in size by appending new values to the old values. Let $\mathbf{x} = [1\ 3\ 5]$:

$$\mathbf{x} = [\mathbf{x}\ 6\ 8\ 10] \implies \mathbf{x} = \begin{bmatrix} 1 & 3 & 5 & 6 & 8 & 10 \end{bmatrix}$$

$$\mathbf{y} = [\mathbf{x}; 1:6] \implies \mathbf{y} = \begin{bmatrix} 1 & 3 & 5 & 6 & 8 & 10 \\ 1 & 2 & 3 & 4 & 5 & 6 \end{bmatrix}.$$

A.9 Graphics

In Chapter 1, you learn that complex variables can be represented as points in the plane. MATLAB makes it easy for you to plot complex variables in a graph. Type `z1 = 1+.5*j;plot(z1,'o')`. The graph window should be activated and the point `z1` displayed by a `'o'`. You must specify the symbol for display, and the authorized symbols for point display are `'.'`, `'o'`, `'+'`, `'x'`, and `'*'`. When you are displaying a curve (to come later), no type is necessary. MATLAB automatically adjusts the scale on a graph to accommodate the value of the point being plotted. In this case, the range is $[0, 2]$ for the real part and $[0, 1]$ for the imaginary part.

Let's now plot a second complex number by typing `z2 = 2+1.5*j; plot(z2,'o')`. Note that the second `plot` command erases the first plot and changes the scaling to $[0, 4]$ and $[0, 3]$. Sometimes you may want to have the points plotted on the same graph. To achieve this, you have to use the command `hold on` after the first plot. Try the following:

```
>> plot(z2,'o')
>> hold on
>> plot(z1,'o')
>> hold off
```

The advantage in using the `hold` command is that there is no limit to the number of plot commands you can type before the hold is turned off, and these plots may involve the same variable plotted over a range of values. You can also use different point displays. A disadvantage of the hold command is that the scaling is enforced by the first plot and is not adjusted for subsequent plots. This is why we plot the point `z2` first. Try reversing the order of the plots and see what happens. This means that points outside the scaling will not be displayed. The command **hold off** permits erasing the current graph for the next plot command.

You can freeze the scaling of the graph by using the command **axis**. MATLAB gives you the message

```
Axis scales frozen
ans =
   0  4  0  3
```

This freezes the current axis scaling for subsequent plots. Similarly, if you type **axis** a second time, MATLAB resumes the automatic scaling feature and prints the message

```
Axis scales auto-ranged
ans =
   0  4  0  3
```

The axis scaling can also be manually enforced by using the command

```
≫ axis([xmin xmax ymin ymax])
```

where (**xmin,ymin**) is the lower left corner and (**xmax,ymax**) is the upper right corner of the graph. This scaling remains in effect until the next **axis** command is entered (with or without arguments).

Another way to plot several complex numbers on the same graph is to display them as a curve. For this purpose, you have to store the numbers in a vector. For example, type **z(1) = z1,z(2) = z2,plot(z)**. Note that the two points are at the two extremes of the line plotted on the graph. If you specify a symbol, then no line is drawn, just the extreme points. Try **plot(z,'o')**.

If you examine your current graph carefully, you will notice that the unit lengths on the x and y axes are not quite the same. In fact, MATLAB adjusts the length of an axis to conform to the overall size of the graph window. What this means is that a 45° line will actually be displayed at an angle depending on the overall aspect ratio of the graph window. To ensure that the aspect ratio is equal to 1, you may enter the command **axis('square')**.

MATLAB will then enforce an aspect ratio equal to 1, regardless of the aspect ratio for the outside graph window. This ensures that circles appear as circles and not as ellipses. MATLAB will make the square graph as large as possible to fit within the graph window. To go back to the default ratio, just type in `axis('normal')`.

To add labels to your graph, the functions `xlabel('text')`, `ylabel('text')`, and `title('text')` are useful and self-explanatory. The argument `text` contains a string of characters. Add the label `Real` on the horizontal axis and the label `Imaginary` on the vertical axis of your graph. The command `grid` draws a grid on your graph. The grid does not remain in effect for the next plots. Try it.

The `plot` Instruction. The `plot` instruction in MATLAB is very versatile. It can be used to plot several different types of data. Its syntax is `plot(x,y,'symbol and/or color')` or `plot(y,'symbol and/or color')`. The instruction will plot a vector of data versus another vector of data. The first vector is referenced to the horizontal axis and the second to the vertical axis. If only one vector is used, then it is plotted with reference to the vertical axis while the horizontal axis is automatically forced to be the index of the vector for the corresponding data point. The notation inside the apostrophes is optionally used to designate whether each element of the vector is to be plotted as a single point with a certain symbol or as a curve with a straight line drawn between each data value. The colors can also be specified. Possible symbols are (*,o,+,.), and colors are (r,g,b,w) (red, green, blue, and white). Complex valued vectors are plotted by making the horizontal axis the real part of the vector and the vertical axis the imaginary part. Warning: A complex valued vector will automatically be plotted correctly on the complex plane (instead of real versus imaginary) *only* if every element of the vector is complex valued. Try `plot (x,y,'*')`, `plot (x,'*')`, `plot (y,'*')`, `plot (y,x,'*')` and `plot (z,'*')`, `plot (Real (z), Imag (z),'*')` to clarify your understanding of plot. Use x = [1 3 5 7], y = [2 4 6 8], z = $[1 + j, 2 + 2j, 3 + 3j]$.

We may summarize as follows:

plot(x,'*r')	(red star—points with the value of x on vertical and indices on horizontal)
plot(y)	(line—connected curve of the value of y on vertical and indices on horizontal)
plot(x,y)	(line—connected curve of the value of y on vertical and the value of x on horizontal)
plot(x,y,'og')	(circle—points of the value of y on vertical and the value of x on horizontal)
plot(real(z),imag(z))	(line—connected plot of z on the complex plane)
plot(real(z),imag(z),'+b')	(blue plus—points of z on the complex plane).

Subplots. It is possible to split the graphics screen up into several separate smaller graphs rather than just one large graph. As many as four subplots can be created. The MATLAB instruction subplot(mnp) signifies which of the smaller graphs is to be accessed with the next plot statement. The mnp argument consists of three digits. The m and n are the numbers of rows (m) and columns (n) into which the screen should be divided. The p designates which of the matrix elements is to be used. For example,

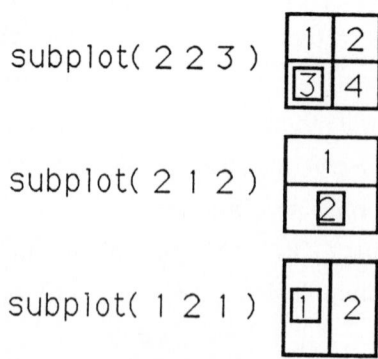

subplot(2 2 3)

subplot(2 1 2)

subplot(1 2 1)

Help and Demos. MATLAB has on-line help and a collection of demonstrations. For a list of available functions, type

```
≫ help
```

For help on a specific function, **sin** for example, type

```
≫ help sin
```

To learn how to use the colon (:, a very important and versatile character) in MATLAB, type

```
≫ help :
```

The demos will also help you become more familiar with MATLAB and its capabilities. To run them, type

```
≫ demo
```

A.10 Editing Files and Creating Functions (Macintosh)

If you quit MATLAB now, all the commands you have typed will be lost. This is where the *Edit* window is useful. If you choose *new* from the *File* menu, a new window appears with the title *Edit 1: Untitled*. In this window, you should type in all the commands you would like MATLAB to execute at once. When you are finished typing, you may save the file by choosing *save* or *save as* in the file menu and by entering a name for the file. If the edit window is active (that is, if it appears in front), then choose *save and go* from the file menu to save the file and execute it. If the command window is active, then you can execute the file by entering its name.

Editing Files. To test your understanding of file editing, enter the following commands in a file named **myfile**:

```
clear,clg                    %Clear variables and graphics
j=sqrt(-1);                  %To be sure
z1=1+.5*j,z2=2+1.5*j         %Enter variables
```

```
z3=z1+z2,z4=z1*z2          %Compute sum and product
axis([0 4 0 4]),           %First plot
   axis('square'),plot(z1,'o')
hold on                    %Allow overplot
plot(z2,'o'),plot(z3,'+'), %Other plots
   plot(z4,'*')
hold off
```

You do not have to type the % sign and the text that follows it. These are simply comments in a file. They are ignored by the MATLAB interpreter. You should, however, make a habit of adding comments (preceded by %) to your file if you want to be able to understand programs that have been written long ago.

Do not forget to save your file. Such a file is called a *script file*. It contains MATLAB commands that could have been entered one by one in the command window. You have three ways to execute a script file:

(i) with the *edit* window active, choose *save and go* from the *file* menu;

(ii) with the *command* window active, enter the file name; or

(iii) with the *command* window active, choose *run script...* from the *M-file* menu. In this case, a menu pops up to ask you which file you want to execute.

Try each of these three methods in order to get used to their differences. Figure A.3 shows the plot that you should get.

Creating Functions. MATLAB puts many commands at your disposal, and you just have to enter their names (with or without arguments) to execute them. Some commands are built in to MATLAB, and others are contained in files to which you have access (not to modify them, but to see how things are done). You can try to display the contents of a command file in the command window by entering the command type filename. Enter type plot. You should get the message

```
??? Built-in function.
```

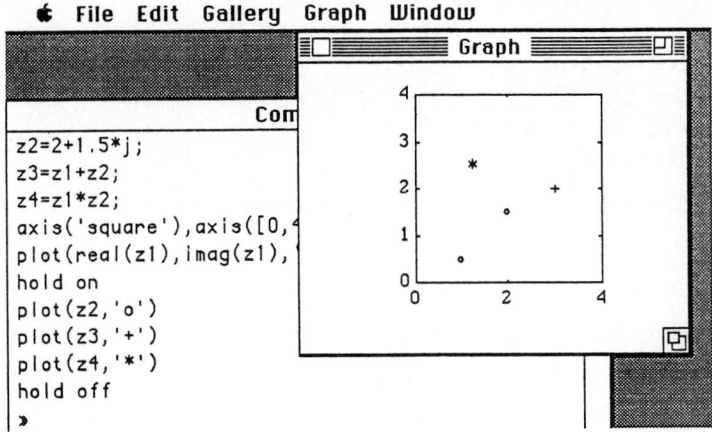

Figure A.3: A Typical Graph (©Apple Computer, Inc., used with permission.)

This means that the command **plot** is a build-in function and thus cannot be typed by the user (it is compiled with the program). Similarly, try to type the function file containing the command **sinh** by entering the command **type sinh**. You should get

```
function y=sinh(x)
%SINH SINH(X) is the hyperbolic sine of the elements of X.
y=(exp(x)-exp(-x))/2;
```

This is a typical example of a *function file*. It differs from a script file by the fact that the first line contains the word **function**. What it does is make the command $y = sinh(x)$ equivalent to the command **y=(exp(x)-exp(-x))/2**. The variable x is the input argument (there could be more than one), and y is the output argument (there also could be more than one). The second line contains comments about the function and its arguments. They can spread to several lines. Display them by typing the command **help sinh**. The third line contains the actual function commands (there could also be several lines).

As an exercise, enter the function perp(x):

```
function y=perp(x)
%PERP PERP(X) is a complex value perpendicular to X.
j=sqrt(-1);
y=j*real(x)-imag(x);
```

Evaluate perp on various complex numbers. Replace the last line by $y = x*j;$. Do you get the same result? Why?

Note that local variables are just that—*local*. Only the input and output arguments are kept in memory after the function is called and executed. For more on functions and multiple arguments, see the MATLAB manual. In the previous example, the variable j is local. If, before using the function perp, you use the same variable j, its contents will not be affected by the command perp(z). Verify this.

Normally, while a file (script or function) is executed, the commands are not displayed on the command window. Just the results are displayed. The command echo allows you to view all the instructions. This is useful for debugging and demonstrations. It is turned off by typing echo off.

A.11 Editing Files and Creating Functions (PC)

You should write a MATLAB program whenever you anticipate executing some sequence of statements several times or again in a later session. On an IBM PC, you may use any text editor to write a program, as long as the file can be saved in ASCII format without the control codes used by most word processors. Applicable text editors include Edix, Wordstar, XTree Pro, and Turbo Pascal's editor.

Editing Files. If you have enough memory, you can run your editor without leaving MATLAB by using the exclamation point (!), like this:

```
≫ !EDIX
```

The exclamation point may be used to execute any DOS command or program from MATLAB. When the command or program finishes, your MATLAB variables are just as you left them. Use your editor to write program lines just as you would type them in MATLAB's command mode. Then save the file with extension .m in the directory where you will run MATLAB. Such MATLAB programs are called m-files. You may run your m-file by typing the file name (without the .m extension) at the MATLAB command prompt (≫).

Script Files. There are two kinds of m-files, called *script files* and *functions*. Running a script file is exactly like typing the commands it contains at the ≫ prompt. Your m-file will automatically be a script file unless you specify otherwise, as described later. Practice by entering, saving, and running `plotsin.m` as listed next:

```
t = -6:.2:6;
y = sin(t);
plot(t,y)
title('SINE')
pause
grid
xlabel('t')
ylabel('sin(t)')
```

When the `pause` is executed, you will need to press a key to go on. If you type `whos` after running `plotsin`, you will see that the variables t and y remain in memory. Comments are important to a script file. They are marked with the symbol %. Anything following this symbol on a line is assumed to be a comment and is ignored by the MATLAB program interpreter.

Functions. Functions differ from script files in that they have designated input and output variables. Any other variables used within a function are local variables, which do not remain after the function terminates and which have no effect on variables outside the function. Many of the functions supplied with MATLAB are actually m-files. A good example is `triu.m`:

```
≫ type triu
```

The word `function` at the beginning of the file makes it a function rather than a script file. The function name in this line must match the file name. The input variables of triu are x and k, meaning that the first input argument will be referred to as x and the second as k within the function. Likewise, the function line designates y as the output. There is nothing special about the variable names x, k, and y when the function is used. It is only that whatever inputs and output you use will be referred to as x, k, and y inside the function. The variables m, n, j, and i are created temporarily when triu runs and disappear when it terminates. They are local variables and have no effect on variables with the same names outside the function. In contrast, a script file has no local variables and does no substitution of input and output variable names.

As an exercise, enter and save the function perp.m:

```
function y=perp(x)
% PERP(x) is a complex number perpendicular to x.
j = sqrt(-1);
y = j*real(x)-imag(x);
```

Evaluate perp on various complex numbers. Replace the last line by y = x*j;. Do you get the same result? Why?

Printing Files and Graphics. To display an m-file on the screen, use the instruction ≫ type filename. To make a copy at a printer, use the DOS command ≫ !print filename.m. Graphics hardcopy is available through the commands meta and gpp. See the MATLAB manual for more information.

A.12 Loops and Control

MATLAB has control statements like those in most computer languages. We will only study the for loop here. See the MATLAB manual for details on if and while statements.

What for loops do is allow a statement or a group of statements to be repeated. For example,

```
for i = 1:n,x(i) = 0,end
```

assigns the value 0 to the first **n** elements of the array **x**. If **n** is less than 1, the instruction is still valid but nothing will be done. If **n** is not defined, then the following message will appear:

```
??? Undefined function or variable.
Symbol in question==>n
```

If **n** contains a real value, the integer part of **n** is used. If **n** is a complex number, the integer part of the real part is taken. (This should, however, be avoided.) If **x** is not declared earlier or is of smaller dimension than **n**, then extra memory is allocated.

The command **end** must follow the command **for**. If it is not present, MATLAB will wait for remaining statements until you type the command **end**, and nothing will be executed in the meantime.

More than one statement may be included in the loop. The statement 1:n is the way to specify all the integer values between 1 and **n**. A step different than 1 may be specified. For example, the statement **for** i=1:2:5,x(i)=1,end is equivalent to x(1)=1,x(3)=1,x(5)=1. Negative steps are also allowed, as in i=n:-1:1.

We may use a **for** loop to draw a circle of radius 1. Type

```
>> j=sqrt(-1);
>> n=360;
>> for i=1:n,circle(i)=exp(2*j*i*pi/n);end;
>> plot(circle)
```

Note how easy it is to plot a curve. But also note how slowly the **for** loop is executed. MATLAB is not very good at executing things one by one. It prefers, by far, a vector-oriented statement. Using the range specification as in the **for** loop, it is possible to speed up the process by replacing the explicit **for** loop by an implicit **for** loop using the colon, like this:

```
>> circle = exp((2*j*pi/n)*[1:n]);
```

Note how much faster this command is executed. In general, **for** loops should be avoided as much as possible. For example, the first **for** loop you typed could have been replaced by the command x=zeros(1,n);, which is much more efficient. The function **zeros** fills the variable with 0's to the specified size. Similarly, the function **ones** fills the variable with 1's. The size can also be determined by the size of the argument. If **A** is a matrix of size **m,n**, then the command **B=ones(A)** fills the matrix B with 1's and forces the matrix B to have exactly the same size as the matrix **A**.

Avoiding for Loops. Since **for** loops are very inefficient in MATLAB (they are computed sequentially, adding several more computations for every loop), it is preferable to use the matrix capabilities of MATLAB to replace **for** loops and speed up processing time.

(i) Replace the **for** loop

```
≫ for i = 1:10,
≫     x(i) = i;
≫ end;
```

with

```
≫ x = 1:10
```

to get

```
x = [1  2  3  4  5  6  7  8  9  10]
```

(ii) Replace the **for** loop

```
≫ z = something;
≫ for i = 1:10,
≫     x(i) = z*i;
≫ end;
```

with

```
≫ z = something;
≫ x = z*(1:10);
```

to get

```
x = [z  2*z  3*z  ...  10*z]
```

(iii) Replace the `for` loop

```
>> z = something;
>> x(1) = z;
>> for i = 2:10,
>>     x(i) = z*x(i-1);
>> end;
```

with

```
>> z = something;
>> x = z.^(1:10);
```

to get

```
x = [z  z**2  z**3  ...  z**10]
```

(iv) Replace the `for` loop

```
>> for i = 0:2:100,
>>     x(i) = 1.0*exp(j*2*pi*i/100);
>> end;
```

with

```
>> x = 1.0*exp(j*2*pi*(0:2:100)/100);
```

to get

```
x = [exp(0)  exp(j*4*pi/100)  ...  exp(j*200*pi/100)]
```

| | | | | | | | | | | # B |

The Edix Editor

On a PC, creating and editing m-files is done outside of MATLAB. The Edix editor is one of the visual types of editors. It can be accessed either from DOS by using > `edix filename.m` or from MATLAB by using ≫ `!edix filename.m`. The latter will return you back to MATLAB after you quit the editor. MATLAB script files must end with a `.m` extension. They are also called m-files. Once in the Edix editor, a help list of the editor's commands can be displayed on the screen by typing "Alt-h." (Note: The "Alt" key must be held down while the "h" key is depressed.) Pressing the "space" key will get you out of the help system.

A new file is created by calling the Edix editor with a file name that has not been previously used. If the file name already exists on disc, then that file is retrieved and displayed on the screen and is ready to be edited. Since Edix is a visual editor, changes are made by moving the cursor to wherever something needs to be typed in or deleted. Once changes have been made to a file, then that file needs to be saved (written to disc) by typing "Alt-w." When all editing is finished, you may exit from the Edix editor by typing "Alt-x." Following are some useful Edix editor commands:

Alt-h(F7) help-list of all the Edix commands;

Alt-w saves (writes) the file to disc;

Alt-x exits the Edix editor;

Alt-d deletes the current line;

Alt-m (F8) the first time used, marks the beginning of a block of lines to be edited; the second time, marks the end of a block of lines to be edited and highlights this block in white; the third time, "un-marks" the highlighted block;

Alt-c makes a copy of the highlighted block at the current cursor location (can be repeated);

Alt-m moves the highlighted block to the current cursor location (can be repeated);

(F5) puts the cursor at the beginning of the line; and

(F6) puts the cursor at the end of the line.

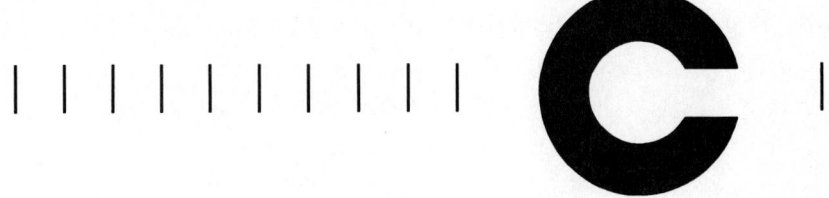

Useful Mathematical Identities

C.1 $e^{j\theta}$, $\cos\theta$, **and** $\sin\theta$

$$e^{j\theta} = \lim_{n\to\infty} \left(1 + j\,\frac{\theta}{n}\right)^n = \sum_{n=0}^{\infty} \frac{1}{n!}\,(j\theta)^n = \cos\theta + j\sin\theta$$

$$\cos\theta = \sum_{n=0}^{\infty} \frac{(-1)^n}{(2n)!}\,\theta^{2n}; \qquad \sin\theta = \sum_{n=0}^{\infty} \frac{(-1)^n}{(2n+1)!}\,\theta^{2n+1}$$

C.2 Trigonometric Identities

$$\sin^2 \theta + \cos^2 \theta = 1$$

$$\sin(\theta + \phi) = \sin \theta \cos \phi + \cos \theta \sin \phi$$

$$\cos(\theta + \phi) = \cos \theta \cos \phi - \sin \theta \sin \phi$$

$$\sin(\theta - \phi) = \sin \theta \cos \phi - \cos \theta \sin \phi$$

$$\cos(\theta - \phi) = \cos \theta \cos \phi + \sin \theta \sin \phi$$

C.3 Euler's Equations

$$e^{j\theta} = \cos \theta + j \sin \theta$$

$$\sin \theta = \frac{e^{j\theta} - e^{-j\theta}}{2j}$$

$$\cos \theta = \frac{e^{j\theta} + e^{-j\theta}}{2}$$

C.4 De Moivre's Identity

$$(\cos \theta + j \sin \theta)^n = \cos n\theta + j \sin n\theta$$

C.5 Binomial Expansion

$$(x + y)^N = \sum_{n=0}^{N} \binom{N}{n} x^n y^{N-n}; \qquad \binom{N}{n} = \frac{N!}{(N - n)!n!}$$

$$2^N = \sum_{n=0}^{N} \binom{N}{n}$$

C.6 Geometric Sums

$$\sum_{k=0}^{\infty} az^k = \frac{a}{1-z}, \ |z| < 1$$

$$\sum_{k=0}^{N-1} az^k = \frac{a(1-z^N)}{1-z}, \ z \neq 1$$

C.7 Taylor's Series

$$f(x) = \sum_{k=0}^{\infty} f^{(k)}(a) \frac{(x-a)^k}{k!}$$

(Maclaurin's Series if $a = 0$)

| | | | | | | | | | | **Index** |